CHILDREN'S RIGHTS

A BOOK OF NURSERY LOGIC

BY

KATE DOUGLAS WIGGIN

" A court as of angels,
  A public not to be bribed,
  Not to be entreated,
  Not to be overawed."

BOSTON AND NEW YORK
HOUGHTON, MIFFLIN AND COMPANY
The Riverside Press, Cambridge
1893

PREFATORY NOTE

I AM indebted to the Editors of Scribner's Magazine, the Cosmopolitan, and Babyhood, for permission to reprint the three essays which have appeared in their pages. The others are published for the first time.

It may be well to ward off the full seriousness of my title "Nursery Logic" by saying that a certain informality in all of these papers arises from the fact that they were originally talks given before members of societies interested in the training of children.

Three of them — "Children's Stories," "How Shall we Govern our Children," and "The Magic of 'Together'" — have been written for this book by my sister, Miss Nora Smith.

K. D. W.

NEW YORK, *August*, 1892.

CONTENTS

THE RIGHTS OF THE CHILD

"Give me liberty, or give me death!"

THE RIGHTS OF THE CHILD

THE subject of Children's Rights does not provoke much sentimentalism in this country, where, as somebody says, the present problem of the children is the painless extinction of their elders. I interviewed the man who washes my windows, the other morning, with the purpose of getting at the level of his mind in the matter.

"Dennis," I said, as he was polishing the glass, "I am writing an article on the 'Rights of Children.' What do you think about it?" Dennis carried his forefinger to his head in search of an idea, for he is not accustomed to having his intelligence so violently assaulted, and after a moment's puzzled thought he said, "What do I think about it, mum? Why, I think we'd ought to give 'em to 'em. But Lor', mum, if we don't, they *take* 'em, so what's the odds?" And as he left the room I thought he looked pained that I should spin words and squander ink on such a topic.

The French dressmaker was my next victim. As she fitted the collar of an effete civilization on my nineteenth century neck, I put the same question I had given to Dennis.

"The rights of the child, madame?" she asked, her scissors poised in air.

"Yes, the rights of the child."

"Is it of the American child, madame?"

"Yes," said I nervously, "of the American child."

"Mon Dieu! he has them!"

This may well lead us to consider rights as opposed to privileges. A multitude of privileges, or rather indulgences, can exist with a total disregard of the child's rights. You remember the man who said he could do without necessities if you would give him luxuries enough. The child might say, "I will forego all my privileges, if you will only give me my rights: a little less sentiment, please, — more justice!" There are women who live in perfect puddles of maternal love, who yet seem incapable of justice; generous to a fault, perhaps, but seldom just.

Who owns the child? If the parent owns him, — mind, body, and soul, we must adopt one line of argument ; if, as a human being, he owns himself, we must adopt another. In my thought the parent is simply a divinely appointed guardian, who acts for his child until he attains what we call the age of discretion, — that highly uncertain period which arrives very late in life with some persons, and not at all with others.

The rights of the parent being almost unlimited, it is a very delicate matter to decide just when and where they infringe upon the rights of the child. There is no standard; the child is the creature of circumstances.

The mother can clothe him in Jaeger wool from head to foot, or keep him in low neck, short sleeves and low stockings, because she thinks it pretty ; she can feed him exclusively on raw beef, or on vegetables, or on cereals ; she can give him milk to drink, or let him sip his father's beer and wine; put him to bed at sundown, or keep him up till midnight ; teach him the catechism and the thirty-nine articles, or tell him there is no God ; she can cram him with facts before he has any appetite or power of assimilation, or she can make a fool of him. She can dose him with old-school remedies, with new-school remedies, or she can let him die without remedies because she does n't believe in the reality of disease. She is quite willing to legislate for his stomach, his mind, his soul, her teachableness, it goes without saying, being generally in inverse proportion to her knowledge; for the arrogance of science is humility compared with the pride of ignorance.

In these matters the child has no rights. The only safeguard is the fact that if parents are absolutely brutal, society steps in, removes the untrustworthy guardian, and appoints another. But society does nothing, can do nothing, with the parent who injures the child's soul, breaks his will, makes him grow up a liar or a coward, or murders his faith. It is not very long since we decided that when a parent brutally abused his child, it could be taken from him and made the ward of the state; the Society for the Prevention of Cruelty to Children is of later date than the Society for the Prevention of Cruelty to Animals. At a distance of a century and a half we can hardly estimate how powerful a blow Rousseau struck for the rights of the child in his educational romance, "Emile." It was a sort of gospel in its day. Rousseau once arrested and exiled, his book burned by the executioner (a few years before he would have been burned with it), his ideas naturally became a craze. Many of the reforms for which he passionately pleaded are so much a part of our modern thought that we do not realize the fact that in those days of routine, pedantry and slavish worship of authority, they were the daring dreams of an enthusiast, the seeming impossible prophecy of a new era. Aristocratic mothers were converts to his theories, and began nursing their children as he commanded them. Great lords began to learn handicrafts; physical exercise came into vogue; everything that Emile did, other people wanted to do.

With all Rousseau's vagaries, oddities, misconceptions, posings, he rescued the individuality of the child and made a tremendous plea for a more natural, a more human education. He succeeded in making people listen where Rabelais and Montaigne had failed; and he inspired other teachers, notably Pestalozzi and Froebel, who knit up his ragged seams of theory, and translated his dreams into possibilities.

Rousseau vindicated to man the right of "Being." Pestalozzi said "Grow!" Froe-

bel, the greatest of the three, cried " Live !
you give bread to men, but I give men to
themselves ! "

The parent whose sole answer to criticism
or remonstrance is " I have a right to do
what I like with my own child ! " is the only
impossible parent. His moral integument
is too thick to be pierced with any shaft
however keen. To him we can only say as
Jacques did to Orlando, " God be with you ;
let 's meet as little as we can."

But most of us dare not take this ground.
We may not philosophize or formulate, we
may not live up to our theories, but we feel
in greater or less degree the responsibility of
calling a human being hither, and the necessity of guarding and guiding, in one way or
another, that which owes its being to us.

We should all agree, if put to the vote,
that a child has a right to be well born.
That was a trenchant speech of Henry Ward
Beecher's on the subject of being " born
again ; " that if he could be born right the
first time he 'd take his chances on the second. " Hereditary rank," says Washington
Irving, " may be a snare and a delusion, but
hereditary virtue is a patent of innate no-
bility which far outshines the blazonry of
heraldry."

Over the unborn our power is almost that of God, and our responsibility, like His toward us; as we acquit ourselves toward them, so let Him deal with us.

Why should we be astonished at the warped, cold, unhappy, suspicious natures we see about us, when we reflect upon the number of unwished-for, unwelcomed children in the world; — children who at best were never loved until they were seen and known, and were often grudged their being from the moment they began to be. I wonder if sometimes a starved, crippled, agonized human body and soul does not cry out, "Why, O man, O woman — why, being what I am, have you suffered me to be?"

Physiologists and psychologists agree that the influences affecting the child begin before birth. At what hour they begin, how far they can be controlled, how far directed and modified, modern science is not assured; but I imagine those months of preparation were given for other reasons than that the cradle and the basket and the wardrobe might be ready; — those long months of supreme patience, when the life-germ is growing from unconscious to conscious being, and when a host of mysterious influences and impulses are being carried silently from mother to child. And if "beauty born of murmuring sound shall pass into" its "face," how much more subtly shall the grave strength of peace, the sunshine of hope and sweet content, thrill the delicate chords of being, and warm the tender seedling into richer life.

Mrs. Stoddard speaks of that sacred passion, maternal love, that "like an orange-tree, buds and blossoms and bears at once." When a true woman puts her finger for the first time into the tiny hand of her baby, and feels that helpless clutch which tightens her very heart-strings, she is born again with the new-born child.

A mother has a sacred claim on the world; even if that claim rest solely on the fact of her motherhood, and not, alas, on any other. Her life may be a cipher, but when the child comes, God writes a figure before it, and gives it value.

Once the child is born, one of his inalienable rights, which we too often deny him, is the right to his childhood.

If we could only keep from untwisting the morning-glory, only be willing to let the sunshine do it! Dickens said real children went out with powder and top-boots; and yet the children of Dickens's time were simple buds compared with the full-blown miracles of conventionality and erudition we raise nowadays.

There is no substitute for a genuine, free, serene, healthy, bread-and-butter childhood. A fine manhood or womanhood can be built on no other foundation; and yet our American homes are so often filled with hurry and worry, our manner of living is so keyed to concert pitch, our plan of existence so complicated, that we drag the babies along in our wake, and force them to our artificial standards, forgetting that " sweet flowers are slow, and weeds make haste."

If we must, or fancy that we must, lead this false, too feverish life, let us at least spare them! By keeping them forever on tiptoe we are in danger of producing an army of conventional little prigs, who know much more than they should about matters which are profitless even to their elders.

In the matter of clothing, we sacrifice children continually to the " Moloch of maternal vanity," as if the demon of dress did not demand our attention, sap our energy, and thwart our activities soon enough at best.

And the right kind of children, before they are spoiled by fine feathers, do detest being " dressed up " beyond a certain point.

A tiny maid of my acquaintance has an elaborate Parisian gown, which is fastened on the side from top to bottom in some mysterious fashion, by a multitude of tiny buttons and cords. It fits the dear little mouse like a glove, and terminates in a collar which is an instrument of torture to a person whose patience has not been developed from year to year by similar trials. The getting of it on is anguish, and as to the getting of it off, I heard her moan to her nurse the other night, as she wriggled her curly head through the too-small exit, " Oh! only God knows how I hate gettin' peeled out o' this dress!"

The spectacle of a small boy whom I meet sometimes in the horse-cars, under the wing of his predestinate idiot of a mother, wrings my very soul. Silk hat, ruffled shirt, silver-buckled shoes, kid gloves, cane, velvet suit, with one two-inch pocket which is an insult to his sex, — how I pity the pathetic little caricature! Not a spot has he to locate a top, or a marble, or a nail, or a string, or a knife, or a cooky, or a nut; but as a bloodless substitute for these necessities of existence, he has a toy watch (that will not go) and an embroidered handkerchief with cologne on it.

As to keeping children too clean for any mortal use, I suppose nothing is more disastrous. The divine right to be gloriously dirty a large portion of the time, when dirt is a necessary consequence of direct, useful, friendly contact with all sorts of interesting, helpful things, is too clear to be denied.

The children who have to think of their clothes before playing with the dogs, digging in the sand, helping the stableman, working in the shed, building a bridge, or weeding the garden, never get half their legitimate enjoyment out of life. And unhappy fate, do not many of us have to bring up children without a vestige of a dog, or a sand heap, or a stable, or a shed, or a brook, or a garden! Conceive, if you can, a more difficult problem than giving a child his rights in a city flat. You may say that neither do we get ours: but bad as we are, we are always good enough to wish for our children the joys we miss ourselves.

Thrice happy is the country child, or the one who can spend a part of his young life among living things, near to Nature's heart.

How blessed is the little toddling thing who can lie flat in the sunshine and drink in the beauty of the " green things growing," who can live among the other little animals, his brothers and sisters in feathers and fur ; who can put his hand in that of dear mother Nature, and learn his first baby lessons without any meddlesome middleman; who is cradled in sweet sounds "from early morn to dewy eve," lulled to his morning nap by hum of crickets and bees, and to his night's slumber by the sighing of the wind, the plash of waves, or the ripple of a river. He is a part of the " shining web of creation," learning to spell out the universe letter by letter as he grows sweetly, serenely, into a knowledge of its laws.

I have a good deal of sympathy for the little people during their first eight or ten years, when they are just beginning to learn life's lessons, and when the laws which govern them must often seem so strange and unjust. It is not an occasion for a big burning sympathy, perhaps, but for a tender little one, with a half smile in it, as we think of what we were, and "what in young clothes we hoped to be, and of how many things have come across; " for childhood is an eternal promise which no man ever keeps.

The child has a right to a place of his own, to things of his own, to surroundings which have some relation to his size, his desires, and his capabilities.

How should we like to live, half the time, in a place where the piano was twelve feet tall, the door knobs at an impossible height, and the mantel shelf in the sky; where every mortal thing was out of reach except a collection of highly interesting objects on dressing-tables and bureaus, guarded, however, by giants and giantesses, three times as large and powerful as ourselves, forever saying, "must n't touch;" and if we did touch we should be spanked, and have no other method of revenge save to spank back symbolically on the inoffensive persons of our dolls?

Things in general are so disproportionate to the child's stature, so far from his organs of prehension, so much above his horizontal line of vision, so much ampler than his immediate surroundings, that there is, between him and all these big things, a gap to be filled only by a microcosm of playthings which give him his first object-lessons. In proof of which let him see a lady richly dressed, he hardly notices her; let him see a doll in similar attire, he will be ravished with ecstasy. As if to show that it was the disproportion of the sizes which unfitted him to notice the lady, the larger he grows the bigger he wants his toys, till, when his wish reaches to life-sizes, good-by to the trumpery, and onward with realities.[1]

My little nephew was prowling about my sitting-room during the absence of his nurse. I was busy writing, and when he took up a delicate pearl opera - glass, I stopped his investigations with the time-honored, "No, no, dear, that's for grown-up people."

"Has n't it got any little-boy end?" he asked wistfully.

That "little-boy end" to things is sometimes just what we fail to give, even when we think we are straining every nerve to surround the child with pleasures. For children really want to do the very same things that we want to do, and yet have constantly to be thwarted for their own good. They would like to share all our pleasures; keep the same hours, eat the same food; but they are met on every side with the seemingly

[1] E. Seguin.

impertinent piece of dogmatism, " It is n't good for little boys," or " It is n't nice for little girls."

Robert Louis Stevenson shows, in his " Child's Garden of Verses," that he is one of the very few people who remember and appreciate this phase of childhood. Could anything be more deliciously real than these verses ?

> " In winter I get up at night,
> And dress by yellow candle light :
> In summer, quite the other way,
> I have to go to bed by day ;
> I have to go to bed and see
> The birds still hopping on the tree,
> And hear the grown-up people's feet
> Still going past me on the street.
> And does it not seem hard to you,
> That when the sky is clear and blue,
> And I should like so much to play,
> I have to go to bed by day ? "

Mr. Hopkinson Smith has written a witty little monograph on this relation of parents and children. I am glad to say, too, that it is addressed to fathers, — that " left wing " of the family guard, which generally manages to retreat during any active engagement, leaving the command to the inferior officer. This " left wing " is imposing on all full-dress parades, but when there is any fighting to be done it retires rapidly to the rear, and only wheels into line when the smoke of the conflict has passed out of the atmosphere.

"Open your heart and your arms wide for your daughters," he says, "and keep them wide open; don't leave all that to their mothers. An intimacy will grow with the years which will fit them for another man's arms and heart when they exchange yours for his. Make a chum of your boy, — hail-fellow-well-met, a comrade. Get down to the level of his boyhood, and bring him gradually up to the level of your manhood. Don't look at him from the second story window of your fatherly superiority and example. Go into the front yard and play ball with him. When he gets into scrapes, don't thrash him as your father did you. Put your arm around his neck, and say you know it is pretty bad, but that he can count on you to help him out, and that you will, every single time, and that if he had let you know earlier, it would have been all the easier."

Again, the child has a right to more justice in his discipline than we are generally wise and patient enough to give him. He is by and by to come in contact with a world where cause and effect follow each other inexorably. He has a right to be taught, and to be governed by the laws under which he must afterwards live ; but in too many cases parents interfere so mischievously and unnecessarily between causes and effects that the child's mind does not, cannot, perceive the logic of things as it should. We might write a pathetic remonstrance against the Decline and Fall of Domestic Authority. There is food for thought, and perhaps for fear, in the subject; but the facts are obvious, and their inevitableness must strike any thoughtful observer of the times. "The old educational régime was akin to the social systems with which it was contemporaneous ; and similarly, in the reverse of these characteristics, our modern modes of culture correspond to our more liberal religious and political institutions."

It is the age of independent criticism. The child problem is merely one phase of the universal problem that confronts society. It seems likely that the rod of reason will have to replace the rod of birch. Parental authority never used to be called into question ; neither was the catechism, nor the

Bible, nor the minister. How should parents hope to escape the universal interrogation point leveled at everything else? In these days of free speech it is hopeless to suppose that even infants can be muzzled. We revel in our republican virtues ; let us accept the vices of those virtues as philosophically as possible.

A lady has been advertising in a New York paper for a German governess "to mind a little girl three years old." The lady's English is doubtless defective, but the fate of the governess is thereby indicated with much greater candor than is usual.

The mother who is most apt to infringe on the rights of her child (of course with the best intentions) is the "firm" person, afflicted with the " lust of dominion." There is no elasticity in her firmness to prevent it from degenerating into obstinacy. It is not the firmness of the tree that bends without breaking, but the firmness of a certain long-eared animal whose force of character has impressed itself on the common mind and become proverbial.

Jean Paul says if " *Pas trop gouverner* " is the best rule in politics, it is equally true of discipline.

But if the child is unhappy who has none of his rights respected, equally wretched is the little despot who has more than his own rights, who has never been taught to respect the rights of others, and whose only conception of the universe is that of an absolute monarchy in which he is sole ruler.

"Children rarely love those who spoil them, and never trust them. Their keen young sense detects the false note in the character and draws its own conclusions, which are generally very just."

The very best theoretical statement of a wise disciplinary method that I know is Herbert Spencer's. "Let the history of your domestic rule typify, in little, the history of our political rule; at the outset, autocratic control, where control is really needful; by and by an incipient constitutionalism, in which the liberty of the subject gains some express recognition; successive extensions of this liberty of the subject; gradually ending in parental abdication."

We must not expect children to be too good; not any better than we ourselves, for example; no, nor even as good. Beware of hothouse virtue. "Already most people recognize the detrimental results of intellectual precocity; but there remains to be recognized the truth that there is a moral precocity which is also detrimental. Our higher moral faculties, like our higher intellectual ones, are comparatively complex. By consequence, they are both comparatively late in their evolution. And with the one as with the other, a very early activity produced by stimulation will be at the expense of the future character."

In these matters the child has a right to expect examples. He lives in the senses; he can only learn through object lessons, can only pass from the concrete example of goodness to a vision of abstract perfection.

"O'er wayward childhood wouldst thou hold firm rule,
And sun thee in the light of happy faces ?
Love, Hope and Patience, these must be thy graces,
And in thine own heart let them first keep school."

Yes, "in thine own heart let them first keep school!" I cannot see why Max O'Rell should have exclaimed with such unction that if he were to be born over again he would choose to be an American woman. He has never tried being one. He does not realize that she not only has in hand the emancipation of the American woman, but the reformation of the American man and the education of the American child. If that triangular mission in life does not keep her out of mischief and make her the angel of the twentieth century, she is a hopeless case.)

Spencer says, " It is a truth yet remaining to be recognized that the last stage in the mental development of each man and woman is to be reached only through the proper discharge of the parental duties. And when this truth is recognized, it will be seen how admirable is the ordination in virtue of which human beings are led by their strongest affections to subject themselves to a discipline which they would else elude."

Women have been fighting many battles for the higher education these last few years; and they have nearly gained the day. When at last complete victory shall perch upon their banners, let them make one more struggle, and that for the highest education, which shall include a specific training for parenthood, a subject thus far quite omitted from the curriculum.

The mistaken idea that instinct is a sufficient guide in so delicate and sacred and vital a matter, the comfortable superstition that babies bring their own directions with them, — these fictions have existed long enough. If a girl asks me why, since the function of parenthood is so uncertain, she should make the sacrifices necessary to such training, sacrifices entailed by this highest education of body, mind, and spirit, I can only say that it is better to be ready, even if one is not called for, than to be called for and found wanting.

CHILDREN'S PLAYS

"The plays of the age are the heart-leaves of the whole future life, for the whole man is visible in them in his finest capacities and his innermost being."

CHILDREN'S PLAYS

Mr. W. W. Newell, in his admirable book on "Children's Games," traces to their proper source all the familiar plays which in one form or another have been handed down from generation to generation, and are still played wherever and whenever children come together in any numbers. The result of his sympathetic and scholarly investigations is most interesting to the student of childhood, and as valuable philologically as historically. In speaking of the old rounds and rhymed formulas which have preserved their vitality under the effacing hand of Time, he says, —

"It will be obvious that many of these well-known game-rhymes were not composed by children. They were formerly played, as in many countries they are still played, by young persons of marriageable age, or even by mature men and women. . . . The truth is, that in past centuries all the world, judged by our present standard, seems to have been a little childish. The maids of honor of Queen Elizabeth's day, if we may credit the poets, were devoted to the game of tag, with which even Diana and her nymphs were supposed to amuse themselves. . . .

"We need not, however, go to remote times or lands for illustration which is supplied by New England country towns of a generation ago. Dancing, under that name, was little practiced; the amusement of young people at their gatherings was "playing games." These games generally resulted in forfeits, to be redeemed by kissing, in every possible variety of position and method. Many of these games were rounds; but as they were not called dances, and as mankind pays more attention to words than things, the religious conscience of the community, which objected to dancing, took no alarm. . . . Such were the pleasures of young men and women from sixteen to twenty-five years of age. Nor were the participants mere rustics; many of them could boast good blood, as careful breeding, and as much intelligence, as any in the land. Neither was the morality or sensitiveness of the young women of that day

in any respect inferior to what it is at
present.

" Now that our country towns are become
mere outlying suburbs of cities, these re-
marks may be read with a smile at the rude
simplicity of old-fashioned American life.
But the laugh should be directed, not at
our own country, but at the bygone age.
It must be remembered that in mediæval
Europe, and in England till the end of the
seventeenth century, a kiss was the usual
salutation of a lady to a gentleman whom
she wished to honor. . . . The Portuguese
ladies who came to England with the In-
fanta in 1662 were not used to the custom ;
but, as Pepys says, in ten days they had
'learnt to kiss and look freely up and
down.' Kissing in games was, therefore, a
matter of course, in all ranks. . . .

" In respectable and cultivated French society, at the time of which we speak, the
amusements, not merely of young people
but of their elders as well, were every whit
as crude.

" Madame Celnart, a recognized a. 'hority
on etiquette, compiled in 1830 a very curious complete manual of society games rec-
ommending them as recreation for *business*

men. . . . 'Their varying movement,' she says, ' their diversity, the gracious and gay ideas which these games inspire, the decorous caresses which they permit, all this combines to give real amusement. These caresses can alarm neither modesty nor prudence, since a kiss in honor given and taken before numerous witnesses is often an act of propriety.' "

The old ballads and nursery rhymes doubtless had much of innocence and freshness in them, but they only come to us nowadays tainted by the odors of city streets. The pleasure and poetry of the original essence are gone, and vulgarity reigns triumphant. If you listen to the words of the games which children play in school yards, on sidewalks, and in the streets on pleasant evenings, you will find that most of them, to say the least, border closely on vulgarity; that they are utterly unsuitable to childhood, notwithstanding that they are played with great glee ; that they are, in fine, common, rude, silly, and boorish. One can never watch a circle of children going through the vulgar inanities of " Jenny O'Jones," " Say, daughter, will you get up ? " " Green Gravel," or " Here come two ducks a-roving," without unspeakable shrinking and moral disgust. These plays are dying out; let them die, for there is a hint of happier things abroad in the air.

The wisest mind of wise antiquity told the riddle of the Sphinx, if having ears to hear we would hear. "Our youth should be educated in a stricter rule from the first, for if education becomes lawless and the youths themselves become lawless, they can never grow up into well-conducted or meritorious citizens; and *the education must begin with, their plays.*"

We talk a great deal about the strength of early impressions. I wonder if we mean all we say; we do not live up to it, at all events. "In childish play deep meaning lies." "The hand that rocks the cradle is the hand that rules the world." "Give me, the first six years of a child's life, and I care not who has the rest." "The child of six years has learned already far more than a student learns in his entire university course." "The first six years are as full of advancement as the six days of creation," and so on. If we did believe these things fully, we should begin education with conscious intelligence at the cradle, if not earlier.

The great German dramatic critic, Schlegel, once sneered at the brothers Jacob and William Grimm, for what he styled their "meditation on the insignificant." These two brothers, says a wiser student, an historian of German literature, were animated by a " pathetic optimism, and possessed that sober imagination which delights in small things and narrow interests, lingering over them with strong affection." They explored villages and hamlets for obscure legends and folk tales, for nursery songs, even ; and bringing to bear on such things at once a human affection and a wise scholarship, their meditation on the insignificant became the basis of their scientific greatness and the source of their popularity. Every child has read some of Grimm's household tales, "The Frog Prince," "Hans in Luck," or the "Two Brothers;" but comparatively few people realize, perhaps, that this collection of stories is the foundation of the modern science of folk-lore, and a by-play in researches of philology and history which place the name of Grimm among the benefactors of our race. I refer to these brothers because they expressed one of the leading theories of the new education.

" My principle," said Jacob Grimm, " has been to undervalue nothing, but to utilize the small for the illustration of the great." When Friedrich Froebel, the founder of the kindergarten, in the course of his researches began to watch the plays of children and to study their unconscious actions, his " meditation on the insignificant " became the basis of scientific greatness, and of an influence still in its infancy, but destined, perhaps, to revolutionize the whole educational method of society.

It was while he was looking on with delight at the plays of little children, their happy, busy plans and make-believes, their intense interest in outward nature, and in putting things together or taking them apart, that Froebel said to himself : " What if we could give the child that which is called education through his voluntary activities, and have him always as eager as he is at play ? "

How well I remember, years ago, the first time I ever joined in a kindergarten game. I was beckoned to the charming circle, and not only one, but a dozen openings were made for me, and immediately, though I was a stranger, a little hand on either side was put into mine, with such friendly, trusting pressure that I felt quite at home. Then we began to sing of the spring-time, and I found myself a green tree waving its branches in the wind. I was frightened and self-conscious, but I did it, and nobody seemed to notice me; then I was a flower opening its petals in the sunshine, and presently, a swallow gathering straws for nest-building; then, carried away by the spirit of the kindergartner and her children, I fluttered my clumsy apologies for wings, and forgetting self, flew about with all the others, as happy as a bird. Soon I found that I, the stranger, had been chosen for the "mother swallow." It was to me, the girl of eighteen, like mounting a throne and being crowned. Four cunning curly heads cuddled under my wings for protection and slumber, and I saw that I was expected to stoop and brood them, which I did, with a feeling of tenderness and responsibility that I had never experienced in my life before. Then, when I followed my baby swallows back to their seats, I saw that the play had broken down every barrier between us, and that they clustered about me as confidingly as if we were old friends. I think I never before felt my own limitations so keenly, or desired so strongly to be fully worthy of a child's trust and love.

Kindergarten play takes the children where they love to be, into the world of " make-believe." In this lovely world the children are, blacksmiths, carpenters, wheelwrights; birds, bees, butterflies; trees, flowers, sunbeams, rainbows; frogs, lambs, ponies, — anything they like. The play is so characteristic, so poetic, so profoundly touching in its simplicity and purity, so full of meaning, that it would inspire us with admiration and respect were it the only salient point of Froebel's educational idea. It endeavors to express the same idea in poetic words, harmonious melody and fitting motion, appealing thus to the thought, feeling, and activity of the child.

Physical impressions are at the beginning of life the only possible medium for awakening the child's sensibility. These impressions should therefore be regulated as systematically as possible, and not left to chance.

Froebel supplies the means for bringing about the result in a simple system of symbolic songs and games, appealing to the child's activities and sensibilities. These, he argues, ought to contain the germ of all later instruction and thought; for physical and sensuous perceptions are the points of departure of all knowledge.

When the child imitates, he begins to understand. Let him imitate the airy flight of the bird, and he enters partially into bird life. Let the little girl personate the hen with her feathery brood of chickens, and her own maternal instinct is quickened, as she guards and guides the wayward motion of the little flock. Let the child play the carpenter, the wheelwright, the wood-sawyer, the farmer, and his intelligence is immediately awakened ; he will see the force, the meaning, the power, and the need of labor. In short, let him mirror in his play all the different aspects of universal life, and his thought will begin to grasp their significance.

Thus kindergarten play may be defined as a " systematized sequence of experiences through which the child grows into self-knowledge, clear observation, and conscious perception of the whole circle of relationships," and the symbols of his play become at length the truth itself, bound fast and deep in heart knowledge, which is deeper and rarer than head knowledge, after all.

To the class occupied exclusively with material things, this phase of Froebel's idea may perhaps seem mystical. There is nothing mystical to children, however; all is real, for their visions have not been dispelled.

> " Turn wheresoe'er I may,
>   By night or day,
> The things which I have seen, I now can see no more."

As soon as the child begins to be conscious of his own activities and his power of regulating them, he desires to imitate the actions of his future life.

Nothing so delights the little girl as to play at housekeeping in her tiny mansion, sacred to the use of dolls. See her whimsical attention to dust and dirt, her tremendous wisdom in dispensing the work and ordering the duties of the household, her careful attention to the morals and manners of her rag-babies.

The boy, too, tries to share in the life of a man, to play at his father's work, to be a miniature carpenter, salesman, or what not. He rides his father's cane and calls it a horse, in the same way that the little girl wraps a shawl about a towel, and showers upon it the tenderest tokens of maternal affection. All these examples go to show that every conscious intellectual phase of the mind has a previous phase in which it was unconscious or merely symbolic.

To get at the spirit and inspiration of symbolic representation in song and game, it is necessary first of all to study Froebel's " Mutter und Kose-Lieder," perhaps the most strikingly original, instructive, serviceable book in the whole history of the practice of education. The significant remark quoted in Froebel's " Reminiscences " is this: " He who understands what I mean by these songs knows my inmost secret." You will find people who say the music in the book is poor, which is largely true, and that the versification is weak, which is often, not always, true, and is sometimes to be attributed to faulty translation; but the idea, the spirit, the continuity of the plan, are matchless, and critics who call it trifling or silly are those who have not the seeing eye nor the understanding heart.

Froebel's wife said of it, —

" A superficial mind does not grasp it,
 A gentle mind does not hate it,
 A coarse mind makes fun of it,
 A thoughtful mind alone tries to get at it."

" Froebel [1] considers it his duty to picture the home as it ought to be, not by writing a book of theories and of rules which are easily forgotten, but by accompanying a mother in her daily rounds through house, garden, and field, and by following her to workshop, market, and church. He does not represent a woman of fashion, but prefers one of humbler station, whom he clothes in the old German housewife style. It may be a small sphere she occupies, but there she is the centre, and she completely fills her place. She rejoices in the dignity of her position as educator of a human being whom she has to bring into harmony with God, nature, and man. She thinks nothing too trifling that concerns her child. She watches, clothes, feeds, and trains it in good habits, and when her darling is asleep, her prayers finish the day. She may not have read much about education, but her sympathy with the child suggests means of doing her duty. Love has made her inventive ; she discovers means of amusement, for play ; she talks and sings, sometimes in poetry and sometimes in prose. From mothers in his circle of relations and friends, Froebel has

[1] Eleonore Heerwart.

learned what a mother can do, and although he had no children of his own, his heart vibrated instinctively with the feelings of a mother's joy, hope, and fear. He did not care about the scorn of others, when he felt he must speak with an almost womanly heart to a mother. His own loss of a mother's tender care made him the more appreciate the importance of a mother's love in early infancy. The mother in his book makes use of all the impressions, influences, and agencies with which the child comes in contact: she protects from evil; she stimulates for good; she places the child in direct communication with nature, because she herself admires its beauties. She has a right feeling towards her neighbors, and to all those on whom she depends. A movement of arms and feet teaches her that the child feels its strength and wants to use it. She helps, she lifts, she teaches; and while playing with her baby's hands and feet she is never at a loss for a song or story.

"The mother also knows that it is necessary to train the senses, because they are the active organs which convey food to the intellect. The ear must hear language, music, the gentle accents and warning voices of father and mother. It must distinguish the sounds of the wind, of the water, and of pet animals.

" The eyesight is directed to objects far and near, as the pigeons flying, the hare running, the light flickering on the wall, the calm beauty of the moon, and the twinkling stars in the dark blue sky."

Of the effect of Froebel's symbolic songs and games, with melodious music and appropriate gesture, kindergartners all speak enthusiastically. They know that —

First : The words suggest thought to the child.

Second : The thought suggests gesture.

Third : The gesture aids in producing the proper feeling.

We all believe thoroughly in the influence of mind on body, the inward working outward, but we are not as ready to see the influence of body on mind. Yet if mind or soul acts upon the body, the external gesture and attitude just as truly react upon the inward feeling. " The soul speaks through the body, and the body in return gives command to the soul." All attitudes mean something, and they all influence the state of mind.

Fourth: The melody begets spiritual impressions.

Fifth: The gestures, feeling, and melody unite in giving a sweet and gentle intercourse, in developing love for labor, home, country, associates, and dumb animals, and in unconsciously directing the intellectual powers.

Learning to sing well is the best possible means of learning to speak well, and the exquisite precision which music gives to kindergarten play destroys all rudeness, and does not in the least rob it of its fun or merriment.

" We cannot tell how early the pleasing sense of musical cadence affects a child. In some children it is blended with the earliest, haziest recollection of life at all, as though they had been literally ' cradled in sweet song ; ' and we may be sure that the hearing of musical sounds and singing in association with others are for the child, as for the adult, powerful influences in awakening sympathetic emotion, and pleasure in associated action."

Who can see the kindergarten games, led by a teacher who has grown into their spirit, and ever forget the joy of the spectacle?

It brings tears to the eyes of any woman who has ever been called mother, or ever hopes to be; and I have seen more than one man retire surreptitiously to wipe away his tears. Is it "that touch of nature which makes the whole world kin"? Is it the perfect self-forgetfulness of the children? Is it a touch of self-pity that the radiant visions of our childhood days have been dispelled, and the years have brought the "inevitable yoke"? Or is it the touching sight of so much happiness contrasted with what we know the home life to be?

Sydney Smith says: "If you make children happy now, you will make them happy twenty years hence by the memory of it;" and we know that virtue kindles at the touch of this joy. "Selfishness, rudeness, and similar weedy growths of school-life or of street-independence cannot grow in such an atmosphere. For joy is as foreign to tumult and destruction, to harshness and selfish disregard of others, as the serene, vernal sky with its refreshing breezes is foreign to the uproar and terrors of the hurricane."

For this kind of ideal play we are indebted to Friedrich Froebel, and if he had left no other legacy to childhood, we should exalt him for it.

If you are skeptical, let me beseech you to join the children in a Free Kindergarten, and play with them. You will be convinced, not through your head, perhaps, but through your heart. I remember converting such a grim female once! You know Henry James says, "Some women are unmarried by choice, and others by chance, but Olive Chancellor was unmarried by every implication of her being." Now, this predestinate spinster acquaintance of mine, well nigh spoiled by years of school-teaching in the wrong spirit, was determined to think kindergarten play simply a piece of nauseating frivolity. She tried her best, but, kept in the circle with the children five successive days, she relaxed so completely that it was with the utmost difficulty that she kept herself from being a butterfly or a bird. It is always so; no one can resist the unconscious happiness of children.

As for the good that comes to grown people from playing with children in this joyous freedom and with this deep earnestness of purpose, it is beyond all imagination. If I had a daughter who was frivolous, or

worldly, or selfish, or cold, or unthoughtful,
— who regarded life as a pleasantry, or fell
into the still more stupid mistake of thinking it not worth living, — I should not (at
first) make her read the Bible, or teach in
the Sunday-school, or call on the minister,
or request the prayers of the congregation,
but I should put her in a good Kindergarten
Training School. No normal young woman
can resist the influence of the study of childhood and the daily life among little children, especially the children of the poor : it
is irresistible.

Oh, these tiny teachers! If we only
learned from them all we might, instead of
feeling ourselves over-wise! I never look
down into the still, clear pool of a child's
innocent, questioning eyes without thinking:
" Dear little one, it must be ' give and take '
between thee and me. I have gained something here in all these years, but thou hast
come from thence more lately than have I ;
thou hast a treasure that the years have
stolen from me — share it with me ! "

Let us endeavor, then, to make the child's
life objective to him. Let us unlock to him
the significance of family, social, and national relationships, so that he may grow

into sympathy with them. He loves the symbol which interprets his nature to himself, and in his eager play, he pictures the life he longs to understand.

If we could make such education continuous, if we could surround the child in his earlier years with such an atmosphere of goodness, beauty, and wisdom, none can doubt that he would unconsciously grow into harmony and union with the All-Good, the All-Beautiful, and the All-Wise.

CHILDREN'S PLAYTHINGS

"Books cannot teach what toys inculcate."

CHILDREN'S PLAYTHINGS

IN the preceding chapter we discussed Froebel's plays, and found that the playful spirit which pervades all the kindergarten exercises must not be regarded as trivial, since it has a philosophic motive and a definite, earnest purpose.

We discussed the meaning of childish play, and deplored the lack of good and worthy national nursery plays. Passing then to Froebel's "Mother-Play," we found that the very heart of his educational idea lies in the book, and that it serves as a guide for mothers whose babies are yet in their arms, as well as for those who have little children of four or five years under their care.

We found that in Froebel's plays the mirror is held up to universal life; that the child in playing them grows into unconscious sympathy with the natural, the human, the divine; that by "playing at" the life he longs to understand, he grows at

last into a conscious realization of its mys-
teries — its truth, its meaning, its dignity,
its purpose.

We found that symbolic play leads the
child from the symbol to the truth symbol-
ized.

We discovered that the carefully chosen
words of the kindergarten songs and games
suggest thought to the child, the thought
suggests gesture, the melody begets spiritual
feeling.

We discussed the relation of body and
mind ; the effect of bodily attitudes on feeling and thought, as well as the moulding of
the body by the indwelling mind.

Froebel's playthings are as significant as
his plays. If you examine the materials he
offers children in his "gifts and occupations," you cannot help seeing that they
meet the child's natural wants in a truly
wonderful manner, and that used in connection with conversations and stories and
games they address and develop his love of
movement and his love of rhythm; his desire
to touch and handle, to play and work (to be
busy), and his curiosity to know; his instincts of construction and comparison, his
fondness for gardening and digging in the

earth ; his social impulse, and finally his religious feeling.

Froebel himself says if his educational materials are found useful, it cannot be because of their exterior, which is as simple as possible, and contains nothing new ; but their worth is to be found exclusively in their application. If you can work out his principles (or better ones still when we find better ones) by other means, pray do it if you prefer ; since the object of the kindergartner is not to make Froebel an *idol*, but an *ideal*. He seems to have found type-forms admirable for awaking the higher senses of the child, and unlike the usual scheme of object lessons, they tell a continued story. When the object-method first burst upon the enraptured sight of the teacher, this list of subjects appeared in a printed catalogue, showing the ground of study in a certain school for six months : —

" *Tea, spiders, apple, hippopotamus, cow, cotton, duck, sugar, rabbits, rice, lighthouse, candle, lead-pencil, pins, tiger, clothing, silver, butter-making, giraffe, onion, soda !* "

Such reckless heterogeneity as this is impossible with Froebel's educational materials, for even if they are given to the child without a single word, they carry something of their own logic with them.

They emphasize the gospel of doing, for Froebel believes in positives in teaching, not negatives; in stimulants, not deterrents. How inexpressibly tiresome is the everlasting "Don't!" in some households. Don't get in the fire, don't play in the water, don't tease the kitty, don't trouble the doggy, don't bother the lady, don't interrupt, don't contradict, don't fidget with your brother, and *don't* worry me now; while perhaps in this whole tirade, not a word has been said of something to do.

Let sleeping faults lie as long as possible while we quietly oust them, little by little, by developing the good qualities. Surely the less we use deterrents the better, since they are often the child's first introduction to what is undesirable or wrong. I am quite sure they have something of that effect on grown people. The telling us not to do, and that we cannot, must not, do a certain thing surrounds it with a momentary fascination. If your enemy suggests that there is a pot of Paris green on the piazza, but you must not take a spoonful and dissolve it in a cup of honey and give it to your maiden aunt who has made her will in your favor, your innocent mind hovers for an instant over the murderous idea.

Froebel's play-materials come to the child when he has entered upon the war-path of getting "something to do." If legitimate means fail, then "let the portcullis fall;" the child must be busy.

The fly on the window-pane will be crushed, the kettle tied to the dog's tail, the curtains cut into snips, the baby's hair shingled, — anything that his untiring hands may not pause an instant, — anything that his chubby legs may take his restless body over a circuit of a hundred miles or so before he is immured in his crib for the night.

The child of four or five years is still interested in objects, in the concrete. He wants to see and to hear, to examine and to work with his hands. How absurd then for us to make him fold his arms and keep his active fingers still; or strive to stupefy him with such an opiate as the alphabet. If we can possess our souls and primers in patience for a while, and feed his senses; if we will let him take in living facts and await the result; that result will be that when he has learned to perceive, compare, and construct, he will desire to learn words, for they tell him what others have seen, thought, and done. This reading and writing, what is it, after all, but the signs for things and thoughts? Logically we must first know things, then thoughts, then their records. The law of human progress is from physical activity to mental power, from a Hercules to a Shakespeare, and it is as true for each unit of humanity as it is for the race.

Everything in Froebel's playthings trains the child to quick, accurate observation. They help children to a fuller vision, they lead them to see. Did you ever think how many people there are who " having eyes, see not " ?

Ruskin says, " Hundreds of people can talk for one who can think, but thousands can think for one who can see. To see clearly is poetry, prophecy, religion, all in one."

A gentleman who is trying to write the biography of a great man complained to me lately, that in consulting a dozen of his friends — men and women who had known him as preacher, orator, reformer, and poet — so few of them had anything character-istic and fine to relate. " What," he said

" is the use of trying to write biography
with such mummies for witnesses! They
would have seen just as much if they had
had nothing but glass eyes in their heads."

What is education good for that does not
teach the mind to observe accurately and
define picturesquely? To get at the essence
of an object and clear away the accompany
ing rubbish, this is the only training that
fits men and women to live with any profit
to themselves or pleasure to others. What
a biographer, for example, or at least what
a witness for some other biographer, was
latent in the little boy who, when told by
his teacher to define a bat, said : " He's a
nasty little mouse, with injy-rubber wings
and shoe-string tail, and bites like the
devil." There was an eye worth having!
Agassiz himself could not have hit off better
the salient characteristics of the little creature in question. Had that remarkable boy
been brought into contact, for five minutes
only, with Julius Cæsar, who can doubt that
the telling description he would have given
of him would have come down through all
the ages ?

I do not mean to urge the adoption of any
ultra-utilitarian standpoint in regard to play-

things, or advise you rudely to enter the realm of early infancy and interfere with the baby's legitimate desires by any meddlesome pedagogic reasoning. Choose his toys, wisely and then leave him alone with them. Leave him to the throng of emotional impressions they will call into being. Remember that they speak to his feelings when his mind is not yet open to reason. The toy at this period is surrounded with a halo of poetry and mystery, and lays hold of the imagination and the heart without awaking vulgar curiosity. Thrice happy age when one can hug one's white woolly lamb to one's bibbed breast, kiss its pink bead eyes in irrational ecstasy, and manipulate the squeak in its foreground without desire to explore the cause thereof!

At this period the well-beloved toy, the dumb sharer of the child's joys and sorrows, becomes the nucleus of a thousand enterprises, each rendered more fascinating by its presence and sympathy. If the toy be a horse, they take imaginary journeys together, and the road is doubly delightful because never traveled alone. If it be a house, the child lives therein a different life for every day in the week; for no monarch alive is so all-powerful as he whose throne is the imagination. Little tin soldier, Shem, Ham, and Japhet from the Noah's Ark, the hornless cow, the tailless dog, and the elephant that won't stand up, these play their allotted parts in his innocent comedies, and meanwhile he grows steadily in sympathy and in comprehension of the ever-widening circle of human relationships. " When we have restored playthings to their place in education — a place which assigns them the principal part in the development of human sympathies, we can later on put in the hands of children objects whose impressions will reach their minds more particularly."

Dr. E. Seguin, our Commissioner of Education to the Universal Exhibition at Vienna, philosophizes most charmingly on children's toys in his Report (chapter on the Training of Special Senses). He says the vast array of playthings (separated by nationalities) left at first sight an impression of silly sameness; but that a second look " discovered in them particular characters, as of national idiosyncrasies; and a closer examination showed that these puerilities had sense enough in them, not only to disclose the movements of the mind, but to predict what is to follow."

He classifies the toys exhibited, and in so doing gives us delightful and valuable generalizations, some of which I will quote: —

" Chinese and Japanese toys innumerable, as was to have been expected. Japanese toys much brighter, the dolls relieved in gold and gaudy colors, absolutely saucy. The application of the natural and mechanical forces in their toys cannot fail to determine the taste of the next generation towards physical sciences.

" Chinese dolls are sober in color, meek in demeanor, and comprehensive in mien. . . . The favorite Chinese toy remains the theatrical scene where the family is treated *à la Molière.*

" Persia sends beautiful toys, from which can be inferred a national taste for music, since most of their dolls are blowing instruments.

" Turkey, Egypt, Arabia, have sent no dolls. Do they make none, under the impression, correct in a low state of culture, that dolls for children become idols for men ?

" The Finlanders and Laplanders, who are not troubled with such religious prejudices, give rosy cheeks and bodies as fat as seals to their dolls.

" The French toy represents the versatility of the nation, touching every topic, grave or grotesque.

" From Berlin come long trains of artillery, regiments of lead, horse and foot on moving tramways.

" From the Hartz and the Alps still issue those wooden herds, more characteristic of the dull feelings of their makers than of the instincts of the animals they represent.

" The American toys justify the rule we have found good elsewhere, that their character both reveals and prefaces the national tendencies. With us, toys refer the mind and habits of children to home economy, husbandry, and mechanical labor; and their very material is durable, mainly wood and iron.

" So from childhood every people has its sympathies expressed or suppressed, and set deeper in its flesh and blood than scholastic ideas. . . . The children who have no toys seize realities very late, and never form ideals. . . . The nations rendered famous by their artists, artisans, and idealists have supplied their infants with many toys, for there is more philosophy and poetry in a single doll than in a thousand books. . . .

If you will tell us what your children play
with, we will tell you what sort of women
and' men they will be; so let this Republic
make the toys which will raise the moral
and artistic character of her children."

Froebel's educational toys do us one service, in that they preach a silent but impres-
sive sermon on simplicity.

It is easy to see that the hurlyburly of
our modern life is not wholly favorable to
the simple creed of childhood, " delight and
liberty, when busy or at rest," but we might
make it a little less artificial than we do, perhaps.

Every thoughtful person knows that the
simple, natural playthings of the old-fashioned child, which are nothing more than
pegs on which he hangs his glowing fancies,
are healthier than our complicated modern
mechanisms, in which the child has only to
" press the button " and the toy " does the
rest."

The electric-talking doll, for example —
imagine a generation of children brought up
on that!   And the toy-makers are not even
content with this grand personage, four feet
high, who says " Papa! Mamma! "   She is
passée already ; they have begun to improve

on her! An electrician described to me the
other day a superb new altruistic doll, fitted
to the needs of the present decade. You
are to press a judiciously located button and
ask her the test question, which is, if she
will have some candy; whereupon with an
angelic detached-movement-smile (located in
the left cheek), she is to answer, " Give
brother *big* piece; give me little piece! "
If the thing gets out of order (and I de-
voutly hope it will), it will doubtless return
to a state of nature, and horrify the by-
standers by remarking, " Give me *big* piece!
Give brother *little* piece! "

Think of having a gilded dummy like
that given you to amuse yourself with!
Think of having to play, — to *play*, for-
sooth, with a model of propriety, a high-
minded monstrosity like that! Does n't it
make you long for your dear old darkey doll
with the raveled mouth, and the stuffing leak-
ing out of her legs; or your beloved Arabella
Clarinda with the broken nose, beautiful
even in dissolution, — creatures " not too
bright or good for human nature's daily
food "? Banged, battered, hairless, sharers
of our mad joys and reckless sorrows, how
we loved them in their simple ugliness!

With what halos of romance we surrounded them! with what devotion we nursed the one with the broken head, in those early days when new heads were not to be bought at the nearest shop.   And even if they could have been purchased for us, would we, the primitive children of those dear, dark ages, have ever thought of wrenching off the cracked blonde head of Ethelinda and buying a new, strange, nameless brunette head, gluing it calmly on Ethelinda's body, as a small acquaintance of mine did last week, apparently without a single pang? Never! A doll had a personality in those times, and has yet, to a few simple backwoods souls, even in this day and generation.   Think of Charles Kingsley's song, — "I once had a sweet little doll, dears."   Can we imagine that as written about one of these modern monstrosities with eyeglasses and corsets and vinaigrettes?

" I once had a sweet little doll, dears,
   The prettiest doll in the world,
   Her face was so red and so white, dears,
   And her hair was so charmingly curled ;
   But I lost my poor little doll, dears,
   As I played on the heath one day,
   And I sought for her more than a week, dears,
   But I never could find where she lay.

" I found my poor little doll, dears,
    As I played on the heath one day ;
Folks say she is terribly changed, dears,
    For her paint is all washed away ;
And her arms trodden off by the cows, dears,
    And her hair not the least bit curled ;
Yet for old sake's sake she is still, dears,
    The prettiest doll in the world."

Long live the doll !

" Dolly-o'diamonds, precious lamb,
Humming-bird, honey-pot, jewel, jam,
Darling delicate-dear-delight —
Angel-o'red, angel-o'white ! "

" Take away the doll, you erase from the heart and head feelings, images, poetry, aspiration, experience, ready for application to real life."

Every mother knows the development of tenderness and motherliness that goes on in her little girl through the nursing and petting and teaching and caring for her doll. There is a good deal of journalistic anxiety concerning the decline of mothers. Is it possible that fathers, too, are in any danger of decline? It is impossible to overestimate the sacredness and importance of the mother-spirit in the universe, but the father-spirit is not positively valueless (so far as it goes). The newspaper-pessimists talk comparatively little about developing that in the young male of the species. In three years' practical experience among the children of the poorer classes, and during all the succeeding years, when I have filled the honorary and honorable offices of general-utility woman, story-teller, song-singer, and playmaker-in-ordinary to their royal highnesses, some thousands of babies, I have been struck with the greater hardness of the small boys; a certain coarseness of fibre and lack of sensitiveness which makes them less susceptible, at first, to gentle influences.

Once upon a time I set about developing this father spirit in a group of little gamins whose general attitude toward the weaker sex, toward birds and flowers and insects, toward beauty in distress and wounded sensibility, was in the last degree offensive. In the bird games we had always had a mother bird in the nest with the birdlings; we now introduced a father bird into the game. Though the children had been only a little time in the kindergarten, and were not fully baptized into the spirit of play, still the boys were generally willing to personate the father bird, since their delight in the active and manly occupation of flying

about the room seeking worms overshadowed
their natural repugnance to feeding the
young. This accomplished, we played " Master Rider," in which a small urchin capered
about on a hobby horse, going through a
variety of adventures, and finally returning
with presents to wife and children. This in
turn became a matter of natural experience,
and we moved towards our grand *coup
d'état*.

Once a week we had dolls' day, when all
the children who owned them brought their
dolls, and the exercises were ordered with
the single view of amusing and edifying
them. The picture of that circle of ragged
children comes before me now and dims my
eyes with its pathetic suggestions.

Such dolls! Five-cent, ten-cent dolls;
dolls with soiled clothes and dolls in a highly
indecorous state of nudity; dolls whose
ruddy hues of health had been absorbed
into their mothers' systems; dolls made of
rags, dolls made of carrots, and dolls made
of towels; but all dispensing odors of garlic
in the common air. Maternal affection,
however, pardoned all limitations, and they
were clasped as fondly to maternal bosoms
as if they had been imported from Paris.

" Bless my soul ! " might have been the un-
spoken comment of these tiny mothers.   " If
we are only to love our offspring when
handsome and well clothed, then the mother-
heart of society is in a bad way ! "

Dolls' day was the day for lullabies.   I
always wished I might gather a group of
stony-hearted men and women in that room
and see them melt under the magic of the
scene.   Perhaps you cannot imagine the
union of garlic and magic, nevertheless, O
ye of little faith, it may exist.   The kindergarten cradle stood in the centre of the
circle, and the kindergarten doll, clean,
beautiful, and well dressed, lay inside the
curtains, waiting to be sung to sleep with the
other dolls.   One little girl after another
would go proudly to the " mother's chair "
and rock the cradle, while the other children
hummed their gentle lullabies.   At this
juncture even the older boys (when the influence of the music had stolen in upon
their senses) would glance from side to side
longingly, as much as to say, —

" O Lord, why didst Thou not make thy
servant a female, that he might dandle one
of these interesting objects without degrada-
tion ! "

In such an hour I suddenly said, "Josephus, will you be the father this time?" and without giving him a second to think, we began our familiar lullaby. The radical nature, the full enormity, of the proposition did not (in that moment of sweet expansion) strike Josephus. He moved towards the cradle, seated himself in the chair, put his foot upon the rocker, and rocked the baby soberly, while my heart sang in triumph. After this the fathers as well as the mothers took part in all family games, and this mighty and much-needed reform had been worked through the magic of a fascinating plaything.

WHAT SHALL CHILDREN READ?

" What we make children love and desire is more important than what we make them learn."

WHAT SHALL CHILDREN READ?

WHEN I was a little girl (oh, six most charming words!) — it is not necessary to name the year, but it was so long ago that children were still reminded that they should be seen and not heard, and also that they could eat what was set before them or go without (two maxims that suggest a hoary antiquity of time not easily measured by the senses), — when I was a little girl, I had the great good fortune to live in a country village.

I believe I always had a taste for books; but I will pass over that early period when I manifested it by carrying them to my mouth, and endeavored to assimilate their contents by the cramming process; and also that later stage, which heralded the dawn of the critical faculty, perhaps, when I tore them in bits and held up the tattered fragments with shouts of derisive laughter. Unlike the critic, no more were given me to

mar; but, like the critic, I had marred a
good many ere my vandal hand was stayed.

As soon as I could read, I had free access
to an excellent medical library, the gloom of
which was brightened by a few shelves of
theological works, bequeathed to the family
by some orthodox ancestor, and tempered by
a volume or two of Blackstone; but outside
of these, which were emphatically not the
stuff my dreams were made of, I can only
remember a certain little walnut bookcase
hanging on the wall of the family sitting-
room.

It had but three shelves, yet all the mys-
teries of love and life and death were in the
score of well-worn volumes that stood there
side by side; and we turned to them, year
after year, with undiminished interest. The
number never seemed small, the stories
never grew tame: when we came to the end
of the third shelf, we simply went back and
began again, — a process all too little known
to latter-day children.

I can see them yet, those rows of shabby
and incongruous volumes, the contents of
which were transferred to our hungry little
brains. Some of them are close at hand
now, and I love their ragged corners, their

dog's-eared pages that show the pressure of childish thumbs, and their dear old backs, broken in my service.

There was a red-covered " Book of Snobs ; " " Vanity Fair " with no cover at all ; " Scottish Chiefs " in crimson ; a brown copy of George Sand's " Teverino ; " and next it a green Bailey's " Festus," which I only attacked when mentally rabid, and a little of which went a surprisingly long way ; and then a maroon " David Copperfield," whose pages were limp with my kisses. (To write a book that a child would kiss! Oh, dear reward ! oh, sweet, sweet fame !)

In one corner — spare me your smiles — was a fat autobiography of P. T. Barnum, given me by a grateful farmer for saving the life of a valuable Jersey calf just as she was on the point of strangling herself. This book so inflamed a naturally ardent imagination, that I was with difficulty dissuaded from entering the arena as a circus manager. Considerations of age or sex had no weight with me, and lack of capital eventually proved the deterrent force. On the shelf above were " Kenilworth," " The Lady of the Lake," and half of " Rob Roy." I have always hesitated to read the other half, for fear that it should not end precisely as I made it end when I was forced, by necessity, to supplement Sir Walter Scott. Then there was "Gulliver's Travels," and if any of the stories seemed difficult to believe, I had only to turn to the maps of Lilliput and Brobdingnag, with the degrees of latitude and longitude duly marked, which always convinced me that everything was fair and aboveboard. Of course, there was a great green and gold Shakespeare, not a properly expurgated edition for female seminaries, either, nor even prose tales from Shakespeare adapted to young readers, but the real thing. We expurgated as we read, child fashion, taking into our sleek little heads all that we could comprehend or apprehend, and unconsciously passing over what might have been hurtful, perhaps, at a later period. I suppose we failed to get a very close conception of Shakespeare's colossal genius, but we did get a tremendous and lasting impression of force and power, life and truth.

When we declaimed certain scenes in an upper chamber with sloping walls and dormer windows, a bed for a throne, a cotton umbrella for a sceptre, our creations were harmless enough. If I remember rightly, our nine-year-old Lady Macbeths and Iagos, Falstaffs and Cleopatras, after they had been dipped in the divine alembic of childish innocence, came out so respectable that they would not have brought the historic " blush to the cheek of youth."

On the shelf above the Shakespeare were a few things presumably better suited to childish tastes, — Hawthorne's " Wonder Book," Kingsley's " Water Babies," Miss Edgeworth's " Rosamond," and the " Arabian Nights."

There were also two little tales given us by a wandering revivalist, who was on a starring tour through the New England villages, " How Gussie Grew in Grace," and " Little Harriet's Work for the Heathen," — melodramatic histories of spiritually perfect and physically feeble children who blessed the world for a season, but died young, enlivened by a few pages devoted to completely vicious and adorable ones who lived to curse the world to a good old age.

Last of all, brought out only on state occasions, was a most seductive edition of that nursery Gaborian, " Who Killed Cock Robin? " with colored illustrations in which the heads of the birds were made to move oracularly, by means of cunningly arranged strips pulled from the bottom of the page. This was a relic of infancy, our first introduction to the literature of plot, counterplot, intrigue, and crime, and the mystery of the murder was very real to us. This book, still in existence, with all the birds headless from over-exertion, is always inextricably associated in my mind with childish woes, as a desire on my part to make the birds wag their heads was always contemporaneous, to a second, with a like desire on my sister's part; and on those rare days when the precious volume was taken down, one of us always donned the penitential nightgown early in the afternoon and supped frugally in bed, while the other feasted gloriously at the family board, never quite happy in her virtue, however, since it separated her from beloved vice in disgrace. That paltry tattered volume, when it confronts me from its safe nook in a bureau drawer, makes my heart beat faster and sets me dreaming! Pray tell me if any book read in your later and wiser years ever brings to your mind such vivid memories, to your lips so lingering a smile, to your eye so ready a tear? True enough, "we could never have loved the earth so well if we had had no childhood in it. . . . What novelty is worth that sweet monotony where everything is known and loved because it is known?"

This autobiographical babble is excusable for one reason only.

It is in remembering what books greatly moved us in earlier days ; what books wakened strong and healthy desires, enlarged the horizon of our understanding, and inspired us to generous action, that we get some clue to the books with which to surround our children ; and a reminiscence of this kind becomes a sort of psychological observation. The moment we realize clearly that the books we read in childhood and youth make a profound impression that can never be repeated later (save in some rare crisis of heart and soul, where a printed page marks an epoch in one's mental or spiritual life), then we become reinforced in our opinion that it makes a deal of difference what children read and how they read it.

Agnes Repplier says: " It is part of the irony of life that our discriminating taste for books should be built up on the ashes of an extinct enjoyment."

A book is such a fact to children, its people are so alive and so heartily loved and hated, its scenes so absolutely real! Prone on the hearth-rug before the fire, or curled in the window seat, they forget everything but the story. The shadows deepen, until they can read no longer; but they do not much care, for the window looks into an enchanted region peopled with brilliant fancies. The old garden is sometimes the Forest of Arden, sometimes the Land of Lilliput, sometimes the Border. The gray rock on the river bank is now the cave of Monte Cristo, and now a castle defended by scores of armed knights who peep one by one from the alder-bushes, while Fair Ellen and the lovely Undine float together on the quiet stream.

For forming a truly admirable literary taste, I cannot indeed say much in favor of my own motley collection of books just mentioned, for I was simply tumbled in among them and left to browse, in accordance with Charles Lamb's whimsical plan for Bridget Elia. More might have been added, and some taken away; but they had in them a world of instruction and illumination which children miss who read too exclusively those books written with rigid determination down to their level, neglecting certain old classics for which we fondly believe there are no substitutes. You cannot always persuade the children of this generation to attack " Robinson Crusoe," and if they do they are too sophisticated to thrill properly when they come to Friday's footsteps in the sand. Think of it, my contemporaries: think of substituting for that intense moment some of the modern " tuppenny " climaxes!

I do not wish to drift into a cheap cynicism, and apotheosize the old days at the expense of the new. We are often inclined to paint the Past with a halo round its head which it never wore when it was the Present. We can reproduce neither the children nor the conditions of fifty or even twenty-five years ago. To-day's children must be fitted for to-day's tasks, educated to answer to-day's questions, equipped to solve to-day's problems; but are we helping them to do this in absolutely the best way? At all events, it is difficult to join in the pæan of gratitude for the tons of children's books that are turned out yearly by parental publishers. If the children of the past did not have quite enough deference paid to their individuality, their likes and dislikes, and if their needs were too often left until the needs of everybody else had been considered, — on the other hand, they were not surfeited with well-meant but ill-directed attentions. If the hay was thrown so high in the rack that they could not pluck a single straw without stretching up for it, why, the hay was generally worth stretching for, and was, perhaps, quite as healthful as the sweet and easily digested nursery porridge which some people adopt as exclusive diet for their darlings nowadays.

Let us look a little at some of the famous children's books of a past generation, and see what was their general style and purpose. Take, for instance, those of Mrs. Barbauld, who may be included in that group of men and women who completely altered the style of teaching and writing for children — Rousseau, de Genlis, the Edgeworths, Jacotot, Froebel, and Diesterweg, all great teachers, — didactic, deadly-dull Mrs. Barbauld, who composed, as one of her biographers tells us, " a considerable number of miscellaneous pieces for the instruction and amusement of young persons, especially females." (Girls were always " young females " in those days ; children were " infants," and stories were "tales.") Who can ever forget those " Early Lessons," written for her adopted son Charles, who appeared in the page sometimes in a state of hopeless ignorance and imbecility, and sometimes clad in the wisdom of the ancients? The use of the offensive phrase " excessively pretty," as applied to a lace tidy by a very tiny female named Lucy, brings down upon her sinful head eleven pages of such moralizing as would only be delivered by a modern mamma on hearing a confession of robbery or murder.

All this does strike us as insufferably didactic, yet we cannot approve the virulence with which Southey and Charles Lamb attacked good Mrs. Barbauld in her old age ; for her purpose was eminently earnest, her views of education healthy and sensible for the time in which she lived, her style polished and admirably quiet, her love for young people indubitably sincere and profound, and her character worthy of all respect and admiration in its dignity, womanliness, and strength. Nevertheless, Charles Lamb exclaims in a whimsical burst of spleen : " ' Goody Two Shoes' is out of print, while Mrs. Barbauld's and Mrs. Trimmer's nonsense lies in piles around. Hang them — the cursed reasoning crew, those blights and blasts of all that is human in man and child."

Miss Edgeworth has what seems to us, in these days, the same overplus of sublime purpose, and, though a much greater writer, is quite as desirous of being instructive, first, last, and all the time, and quite as unable or unwilling to veil her purpose. No books, however, have ever had a more remarkable influence upon young people, and there are many of them — old-fashioned as they are — which the sophisticated children of to-day could read with pleasure and profit.

Poor, naughty Rosamond! choosing the immortal " purple jar " out of that apothecary's window, instead of the shoes she needed ; and in a following chapter, after pages of excellent maternal advice, taking the hideous but useful " red morocco housewife " instead of the coveted " plum."

People may say what they like of Miss Edgeworth's lack of proportion as a moralist and economist, but we have few writers for children at present who possess the practical knowledge, mental vigor, and moral force which made her an imposing figure in juvenile literature for nearly a century.

There has never been a time when the difficulty of making a good use of books was as great as it is to-day, or a time when it required so much decision to make a wise choice, simply because there is so much printed matter precipitated upon us that we cannot " see the wood for the trees."

It is not my province to discriminate between the various writers for children at the present time. To give a complete catalogue of useful books for children would be quite impossible; to give a partial list, or endeavor to point out what is worthy and what unworthy, would be little better. No course of reading laid down by one person ever suits another, and the published "lists of best books," with their solemn platitudes in the way of advice, are generally interesting only in their reflection of the writer's personality.

I would not choose too absolutely for a child save in his earliest years, but would rather surround him with the best and worthiest books, and let him choose for himself; for there are elective affinities and antipathies here that need not be disregarded, —

that are, indeed, certain indications of latent powers, and trustworthy guides to the child's unfolding possibilities.

" Books can only be profoundly influential as they unite themselves with decisive tendencies." Provide the right conditions for mental growth, and then let the child do the growing. If we dictate too absolutely, we *en*velop instead of *de*veloping his mind, and weaken his power of choice. On the other hand, we do not wish his reading to be partial or one-sided, as it may be without intelligent oversight.

I was telling bedtime stories, the other night, to a proper, wise, dull little girl of ten years. When I had successfully introduced a mother-cat and kittens to her attention, I plunged into what I thought a graphic and perfectly natural conversation between them, when she cut me short with the observation that she disliked stories in which animals talked, because they were not true! I was rebuked, and tried again with better success, until there came an unlucky figure of speech concerning a blossoming locust-tree, that bent its green boughs and laughed in the summer sunshine, because its flowers were fragrant and lovely, and the world so green and beautiful. This she thought, on sober second thought, a trifle silly, as trees never did laugh! Now, that exasperating scrap of humanity (she is abnormal, to be sure) ought to be locked up and fed upon fairy tales until she is able to catch a faint glimpse of " the light that never was on sea or land." Poor, blind, deaf little person, predestined, perhaps, to be the mother of a lot of other blind, deaf little persons some day, — how I should like to develop her imagination!

Whatever children read, let us see that it is good of its kind, and that it gives variety, so that no integral want of human nature shall be neglected, — so that neither imagination, memory, nor reflection shall be starved. I own it is difficult to help them in their choice, when most of us have not learned to choose wisely for ourselves. A discriminating taste in literature is not to be gained without effort, and our constant reading of the little books spoils our appetite for the great ones.

Style is a matter of some moment, even at this early stage. Mothers sometimes forget that children cannot read slipshod, awkward, redundant prose, and sing-song vapid verse, for ten or twelve years, and then take kindly to the best things afterward.

Long before a child is conscious of such a thing as purity, delicacy, directness, or strength of style, he has been acted upon unconsciously, so that when the period of conscious choice comes, he is either attracted or repelled by what is good, according to his training. Children are fond of vivacity and color, and love a bit of word painting or graceful nonsense; but there are people who strive for this, and miss, after all, the true warmth and geniality that is most desirable for little people. Apropos of nonsense, we remember Leigh Hunt, who says that there are two kinds of nonsense, one resulting from a superabundance of ideas, the other from a want of them. Style in the hands of some writers is like war-paint to the savage — of no perceptible value unless it is laid on thick. Our little ones begin too often on cheap and tawdry stories in one or two syllables, where pictures in primary colors try their best to atone for lack of matter. Then they enter on a prolonged series of children's books, some of them written by people who have neither the intelligence nor the literary skill to write for a more critical audience; on the same basis of reasoning which puts the young and inexperienced teachers into the lowest grades, where the mind ought to be formed, and assigns to the more practiced the simpler task of *inform-* ing the already partially formed (or *de-* formed) mind.

There has never been such conscientious, intelligent, and purposeful work done for children as in the last ten years; and if an overwhelming flood of trash has been poured into our laps along with the better things, we must accept the inevitable. The legends, myths, and fables of the world, as well as its history and romance, are being brought within reach of young readers by writers of wide knowledge and trained skill.

Knowing, then, as we do, the dangers and obstacles in the way, and realizing the innu- merable inspirations which the best thought gives to us, can we not so direct the reading of our children that our older boys and girls shall not be so exclusively modern in their tastes ; so that they may be inclined to take a little less Mr. Saltus, a little more Shake- speare, temper their devotion to Mr. Kipling by small doses of Dante, forsake " The Duchess " for a dip into Thackeray, and use

Hawthorne as a safe and agreeable antidote to Mr. Haggard? We need not despair of the child who does not care to read, for books are not the only means of culture; but they are a very great means when the mind is really stimulated, and not simply padded with them.

Mr. Frederic Harrison says: " Books are no more education than laws are virtue. Of all men, perhaps the book-lover needs most to be reminded that man's business here is. to know for the sake of living, not to live for the sake of knowing."

But a child who has no taste for reading, who is utterly incapable of losing himself in a printed page, quite unable to forget his childish griefs,

> " And plunge,
> Soul forward, headlong into a book's profound,
> Impassioned for its beauty and salt of truth,"

— such a child is to be pitied as missing one of the chief joys of life. Such a child has no dear old book-friendships to look back upon. He has no sweet associations with certain musty covers and time-worn pages; no sacred memories of quiet moments when a new love of goodness, a new throb of generosity, a new sense of humanity, were born

in the ardent young soul; born when we had
turned the last page of some well-thumbed
volume and pressed our tear-stained childish
cheek against the window pane, when it was
growing dusk without, and a mother's voice
called us from our shelter to "Lay the
book down, dear, and come to tea." For,
to speak in better words than my own, " It
is the books we read before middle life that
do most to mould our characters and influence our lives; and this not only because
our natures are then plastic and our opin-
ions flexible, but also because, to produce
lasting impression, it is necessary to give a
great author time and meditation. The
books that are with us in the leisure of
youth, that we love for a time not only
with the enthusiasm, but with something of
the exclusiveness, of a first love, are those
that enter as factors forever in our mental
life."

CHILDREN'S STORIES

"To be a good story-teller is to be a king among children."

CHILDREN'S STORIES

THE business of story-telling is carried on from the soundest of economic motives, in order to supply a constant and growing demand. We are forced to satisfy the clamorous nursery-folk that beset us on every hand.

Beside us stands an eager little creature quivering with expectation, gazing with wide-open eyes, and saying appealingly, " Tell me a story ! " or perhaps a circle of toddlers is gathered round, each one offering the same fervent prayer, with so much trust and confidence expressed in look and gesture that none but a barbarian could bear to disappoint it.

The story-teller is the children's special property. When once his gifts have been found out, he may bid good-by to his quiet snooze by the fire, or his peaceful rest with a favorite book. Though he hide in the uttermost parts of the house, yet will he be discovered and made to deliver up his treasure. On this one subject, at least, the little ones of the earth are a solid, unanimous body; for never yet was seen the child who did not love the story and prize the story-teller.

Perhaps we never dreamed of practicing the art of story-telling till we were drawn into it by the imperious commands of the little ones about us. It is an untrodden path to us, and we scarcely understand as yet its difficulties and hindrances, its breadth and its possibilities. Yet this eager, unceasing demand of the child-nature we must learn to supply, and supply wisely; for we must not think that all the food we give the little one will be sure to agree with him because he is so hungry. This would be no more true of a mental than of a physical diet.

What objects, then, shall our stories serve beyond the important one of pleasing the little listeners? How can we make them distinctly serviceable, filling the difficult and well-nigh impossible *rôle* of "useful as well as ornamental"?

There are, of course, certain general benefits which the child gains in the hearing of all well-told stories. These are, familiarity with good English, cultivation of the imagi-

nation, development of sympathy, and clear
impression of moral truth. We shall find,
however, that all stories appropriate for
young children naturally divide themselves
into the following classes: —

I. The purely imaginative or fanciful,
and·here belongs the so-called fairy story.

II. The realistic, devoted to things which
have happened, and might, could, would, or
should happen without violence to proba-
bility. These are generally the vehicle for
moral lessons which are all the more impres-
sive because not insisted on.

III. The scientific, conveying bits of in-
formàtion about animals, flowers, rocks, and
stars.

IV. The historical, or simple, interesting
accounts of the lives of heroes and events in
our country's struggle for life and liberty.

There is a great difference in opinion
regarding the advisability of telling fairy
stories to very young children, and there
can be no question that some of them are entirely undesirable and inappropriate. Those
containing a fierce or horrible element must,
of course, be promptly ruled out of court,
including the "bluggy" tales of cruel step-
mothers, ferocious giants and ogres, which

fill the so-called fairy literature. Yet those which are pure in tone and gay with fanciful coloring may surely be told occasionally, if only for the quickening of the imagination. Perhaps, however, it is best to keep them as a sort of sweetmeat, to be taken on high days and holidays only.

Let us be realistic, by all means; but beware, O story-teller! of being too realistic. Avoid the " shuddering tale " of the wicked boy who stoned the birds, lest some hearer be inspired to try the dreadful experiment and see if it really does kill. Tell not the story of the bears who were set on a hot stove to learn to dance, for children quickly learn to gloat over the horrible.

Deal with the positive rather than the negative in story-telling; learn to affirm, not to deny.

Some one perhaps will say here, the knowledge of cruelty and sin must come some time to the child; then why shield him from it now? True, it must come; but take heed that you be not the one to introduce it arbitrarily. " Stand far off from childhood," says Jean Paul, " and brush not away the flower-dust with your rough fist."

The truths of botany, of mineralogy, of

zoölogy, may be woven into attractive stories which will prove as interesting to the child as the most extravagant fairy tale. But endeavor to shape your narrative so dexterously around the bit of knowledge you wish to convey, that it may be the pivotal point of interest, that the child may not suspect for a moment your intention of instructing him under the guise of amusement. Should this dark suspicion cross his mind, your power is weakened from that moment, and he will look upon you henceforth as a deeply dyed hypocrite.

The historic story is easily told, and universally interesting, if you make it sufficiently clear and simple. The account of the first Thanksgiving Day, of the discovery of America, of the origin of Independence Day, of the boyhood of our nation's heroes, — all these can be made intelligible and charming to children. I suggest topics dealing with our own country only, because the child must learn to know the near-at-hand before he can appreciate the remote. It is best that he should gain some idea of the growth of his own traditions before he wanders into the history of other lands.

In any story which has to do with soldiers

and battles, do not be too martial. Do not permeate your tale with the roar of guns, the smell of powder, and the cries of the wounded. Inculcate as much as possible the idea of a struggle for a principle, and omit the horrors of war.

We must remember that upon the kind of stories we tell the child depends much of his later taste in literature. We can easily create a hunger for highly spiced and sensational writing by telling grotesque and horrible tales in childhood. When the little one has learned to read, when he holds the key to the mystery of books, then he will seek in them the same food which so gratified his palate in earlier years.

We are just beginning to realize the importance of beginnings in education.

True, a king of Israel whose wisdom is greatly extolled, and whose writings are widely read, urged the importance of the early training of children about three thousand years ago; but the progress of truth in the world is proverbially slow. When parents and teachers, legislators and lawgivers, are at last heartily convinced of the inestimable importance of the first six years of childhood, then the plays and occupations of that formative period of life will no longer be neglected or left to chance, and the exercise of story-telling will assume its proper place as an educative influence.

Long ago, when I was just beginning the study of childhood, and when all its possibilities were rising before me, " up, up, from glory to glory," — long ago, I was asked to give what I considered the qualifications of an ideal kindergartner.

My answer was as follows, — brief perhaps, but certainly comprehensive : —

The music of St. Cecilia.

The art of Raphael.

The dramatic genius of Rachel.

The administrative ability of Cromwell.

The wisdom of Solomon.

The meekness of Moses, and —

The patience of Job.

Twelve years' experience with children has not lowered my ideals one whit, nor led me to deem superfluous any of these qualifications; in fact, I should make the list a little longer were I to write it now, and should add, perhaps, the prudence of Franklin, the inventive power of Edison, and the talent for improvisation of the early Troubadours.

The Troubadours, indeed, could they return to the earth, would wander about lonely and unwelcomed till they found home and refuge in the hospitable atmosphere of the kindergarten, — the only spot in the busy modern world where delighted audiences still gather around the professional story-teller.

If I were asked to furnish a recipe for one of these professional story-tellers, these spinners of childish narratives, I should suggest one measure of pure literary taste, two of gesture and illustration, three of dramatic fire, and four of ready speech and clear expression. If to these you add a pinch of tact and sympathy, the compound should be a toothsome one, and certain to agree with all who taste it.

And now as to the kind of story our professional is to tell. In selecting this, the first point to consider is its suitability to the audience. A story for very little ones, three or four years old perhaps, must be simple, bright, and full of action. They do not yet know how to listen; their comprehension of language is very limited, and their sympathies quite undeveloped. Nor are they prepared to take wing with you into the lofty realms of the imagination : the adventures of the playful kitten, of the birdling learning to fly, of the lost ball, of the faithful dog, — things which lie within their experience and belong to the sweet, familiar atmosphere of the household, — these they enjoy and understand.

It will be found also that the number of children to whom one is talking is a prominent factor in the problem of selecting a story. Two or three little ones, gathered close about you, may pay strict attention to a quiet, calm, eventless history ; but a circle of twenty or thirty eager, restless little people needs more sparkle and incident.

If one is addressing a large number of children, the homes from which they come must be considered. Children of refined, cultivated parents, who have listened to family conversation, who have been talked to and encouraged to express themselves, — these are able to understand much more lofty themes than the poor little mites who are only familiar with plain, practical ideas, and rough speech confined to the most ordinary wants of life.

And now, after the story is well selected, how long shall it be ? It is impossible to

fix an exact limit to the time it should oc-
cupy, for much depends on the age and the
number of the children. I am reminded
again of recipes, and of the dismay of the
inexperienced cook when she reads, " Stir in
flour enough to make a stiff batter." Alas!
how is she who has never made a stiff batter to settle the exact amount of flour necessary?

I might give certain suggestions as to
time, such as, " Close while the interest is
still fresh; " or, " Do not make the tale so
long as to weary the children;" but after
all, these are only cook-book directions. In
this, as in many other departments of work
with children, one must learn in that "dear
school" which "experience keeps." Five
minutes, however, is quite long enough with
the babies, and you will find that twice this
time spent with the older children will give
room for a tale of absorbing interest, with
appropriate introduction and artistic *dénoue-
ment*.

As one of the chief values of the exercise
is the familiarity with good English which
it gives, I need not say that especial attention must be paid to the phraseology in
which the story is clothed. Many persons

who never write ungrammatically are inaccurate in speech, and the very familiarity and ease of manner which the story-teller must assume may lead her into colloquialisms and careless expressions. Of course, however, the language must be simple ; the words, for the most part, Saxon. No ponderous, Johnsonian expressions should drag their slow length through the recital, entangling in their folds the comprehension of the child ; nor, on the other hand, need we confine ourselves to monosyllables, adopting the bald style of Primers and First Readers. It is quite possible to talk simply and yet with grace and feeling, and we may be sure that children invariably appreciate poetry of expression.

The story should always be accompanied with gestures, — simple, free, unstudied motions, descriptive, perhaps, of the sweep of the mother bird's wings as she soars away from the nest, or the waving of the fir-tree's branches as he sings to himself in the sunshine. This universal language is understood at once by the children, and not only serves as an interpreter of words and ideas, but gives life and attraction to the exercise.

Illustrations, either impromptu or carefully prepared beforehand, are always hailed with delight by the children. Nor need you hesitate to try your " 'prentice hand " at this work. Never mind if you " cannot draw." It must be a rude picture, indeed, which is not enjoyed by an audience of little people. Their vivid imaginations will triumph over all difficulties, and enable them to see the ideal shining through the real. It is well now and then, also, to have the children illustrate the story. Their drawings, if executed quite without help, are most interesting from a psychological standpoint, and will afford great delight to you, as well as to the little artists themselves.

The stories can also be illustrated with clay modeling, an idealized mud-pie-making very dear to children. They soon become quite expert in moulding simple objects, and enjoy the work with all the capacity of their childish hearts.

Now and then encourage the little ones to repeat what they remember of the tale you have told, or to tell something new on the same theme. If the story you have given has been within their range and on a familiar subject, a torrent of infantile reminiscence will immediately gush forth, and you will have a miniature "experience meeting." If you have been telling a dog story, for instance, — "I hed a dog once't," cries Jimmy breathlessly, and is just about to tell some startling incident concerning him, when Nickey pipes up, "And so hed I, and the pound man tuk him;" and so on, all around the circle in the Free Kindergarten, each child palpitating with eagerness to give you his bit of personal experience.

Gather the little ones as near to you as possible when you are telling stories, the tiniest in your lap, the others cuddled at your knee. This is easily managed in the nursery, but is more difficult with a large circle of children. With the latter you can but seat yourself among the wee ones, confident that the interest of the story will hold the attention of the older children.

What a happy hour it is, this one of story-telling, dear and sacred to every child-lover! What an eager, delightful audience are these little ones, grieving at the sorrows of the heroes, laughing at their happy successes, breathless with anxiety lest the cat catch the disobedient mouse, clapping hands when the Ugly Duckling is changed into

the Swan, — all appreciation, all interest, all
joy ! We might count the rest of the world
well lost, could we ever be surrounded by
such blooming faces, such loving hearts, and
such ready sympathy.

THE RELATION OF THE KINDER-
GARTEN TO SOCIAL REFORM

"New social and individual wants demand new solu-
tions of the problem of education."

THE RELATION OF THE KINDER-GARTEN TO SOCIAL REFORM

"Social reform!" It is always rather an awe-striking phrase. It seems as if one ought to be a philosopher, even to approach so august a subject. The kindergarten — a simple unpretentious place, where a lot of tiny children work and play together; a place into which if the hard-headed man of business chanced to glance, and if he did not stay long enough, or come often enough, would conclude that the children were frittering away their time, particularly if that same good man of business had weighed and measured and calculated so long that he had lost the seeing eye and understanding heart.

Some years ago, a San Francisco kindergartner was threading her way through a dirty alley, making friendly visits to the children of her flock. As she lingered on a certain door-step, receiving the last confidences of some weary woman's heart, she heard a loud but not unfriendly voice ringing from an upper window of a tenement-house just round the corner. " Clear things from under foot!" pealed the voice, in stentorian accents. " The teacher o' the *Kids' Guards* is comin' down the street!"

" Eureka!" thought the teacher, with a smile. " There 's a bit of sympathetic translation for you! At last, the German word has been put into the vernacular. The odd, foreign syllables have been taken to the ignorant mother by the lisping child, and the *kindergartners* have become the *Kids' Guards!* Heaven bless the rough translation, colloquial as it is! No royal accolade could be dearer to its recipients than this quaint, new christening!"

What has the kindergarten to do with social reform? What bearing have its theory and practice upon the conduct of life?

A brass-buttoned guardian of the peace remarked to a gentleman on a street-corner, " If we could open more kindergartens, sir, we could almost shut up the penitentiaries, sir!" We heard the sentiment, applauded it, and promptly printed it on the cover of three thousand reports; but on calm reflection it appears like an exaggerated statement. I am not sure that a kindergarten in every ward of every city in America " would almost shut up the penitentiaries, sir ! " The most determined optimist is weighed down by the feeling that it will take more than the ardent prosecution of any one reform, however vital, to produce such a result. We appoint investigating committees, who ask more and more questions, compile more and more statistics, and get more and more confused every year. " Are our criminals native or foreign born ? " that we may know whether we are worse or better than other people ? " Have they ever learned a trade ? " that we may prove what we already know, that idle fingers are the devil's tools; " Have they been educated ? " — by any one of the sorry methods that take shelter under that much-abused word, — that we may know whether ignorance is a bliss or a *blister ;* " Are they married or single ? " that we may determine the influence of home ties; " Have they been given to the use of liquor ? " that we may heap proof on proof, mountain high, against the monster evil of intemperance ; " What has been their family history ? " that we may know how heavily the law of heredity has laid its burdens upon them.

Burning questions all, if we would find out the causes of crime.

To discover the why and wherefore of things is a law of human thought. The reform schools, penitentiaries, prisons, insane asylums, hospitals, and poorhouses are all filled to overflowing; and it is entirely sensible to inquire how the people came there, and to relieve, pardon, bless, cure, or reform them as far as we can. Meanwhile, as we are dismissing or blessing or burying the unfortunates from the imposing front gates of our institutions, new throngs are crowding in at the little back doors. Life is a bridge, full of gaping holes, over which we must all travel! A thousand evils of human misery and wickedness flow in a dark current beneath; and the blind, the weak, the stupid, and the reckless are continually falling through into the rushing flood. We must, it is true, organize our life-boats. It is our duty to pluck out the drowning wretches, receive their vows of penitence and gratitude, and pray for courage and resignation when they celebrate their rescue by falling in again. But we agree nowadays that we should do them much better service if we could contrive to mend more of the holes in the bridge.

The kindergarten is trying to mend one of these "holes." It is a tiny one, only large enough for a child's foot; but that is our bit of the world's work, — to *keep it small!* If we can prevent the little people from stumbling, we may hope that the grown folks will have a surer foot and a steadier gait.

A wealthy lady announced her intention of giving $25,000 to some Home for Incurables. "Why," cried a bright kindergartner, "*don't* you give twelve and a half thousand to some Home for *Curables*, and then your other twelve and a half will go so much further?"

In a word, solicitude for childhood is one of the signs of a growing civilization. "To cure, is the voice of the past; to prevent, the divine whisper of to-day."

What is the true relation of the kindergarten to social reform? Evidently, it can have no other relation than that which grows out of its existence as a plan of education. Education, we have all glibly agreed, lessens the prevalence of crime. That sounds very well; but, as a matter of fact, has our past system produced all the results in this direction that we have hoped and prayed for?

The truth is, people will not be made much better by education until the plan of educating them is made better to begin with.

Froebel's idea — the kindergarten idea — of the child and its powers, of humanity and its destiny, of the universe, of the whole problem of living, is somewhat different from that held by the vast majority of parents and teachers. It is imperfectly carried out, even in the kindergarten itself, where a conscious effort is made, and is infrequently attempted in the school or family.

His plan of education covers the entire period between the nursery and the university, and contains certain essential features which bear close relation to the gravest problems of the day. If they could be made an integral part of all our teaching in families, schools, and institutions, the burdens under which society is groaning to-day would fall more and more lightly on each succeeding generation. These essential features have often been enumerated. I am no fortunate herald of new truth. I may not even put the old wine in new bottles; but iteration is next to inspiration, and I shall give you the result of eleven years' experience among the children and homes of the poorer classes. This experience has not been confined to teaching. One does not live among these people day after day, pleading for a welcome for unwished-for babies, standing beside tiny graves, receiving pathetic confidences from wretched fathers and helpless mothers, without facing every problem of this workaday world ; they cannot all be solved, even by the wisest of us; we can only seize the end of the skein nearest to our hand, and patiently endeavor to straighten the tangled threads.

The kindergarten starts out plainly with the assumption that the moral aim in education is the absolute one, and that all others are purely relative. It endeavors to be a life-school, where all the practices of complete living are made a matter of daily habit. It asserts boldly that doing right would not be such an enormously difficult matter if we practiced it a little, — say a tenth as much as we practice the piano, — and it intends to give children plenty of opportunity for practice in this direction. It says insistently and eternally, " Do noble things, not dream them all day long." For development, action is the indispensable requisite. To develop moral feeling and the power and habit of moral doing we must exercise them, excite, encourage, and guide their action. To check, reprove, and punish wrong feeling and doing, however necessary it be for the safety and harmony, nay, for the very existence of any social state, does not develop right feeling and good doing. It does not develop anything, for it stops action, and without action there is no development. At best it stops wrong development, that is all.

In the kindergarten, the physical, mental, and spiritual being is consciously addressed at one and the same time. There is no "piece-work" tolerated. The child is viewed in his threefold relations, as the child of Nature, the child of Man, and the child of God; there is to be no disregarding any one of these divinely appointed relations. It endeavors with equal solicitude to instill correct and logical habits of thought, true and generous habits of feeling, and pure and lofty habits of action; and it asserts serenely that, if information cannot be gained in the right way, it would better not be gained at all. It has no special hobby, unless you would call its eternal plea for the all-sided development of the child a hobby.

Somebody said lately that the kindergarten people had a certain stock of metaphysical statements to be aired on every occasion, and that they were over-fond of prating about the " being " of the child. It would hardly seem as if too much could be said in favor of the symmetrical growth of the child's nature. These are not mere " silken phrases ; " but, if any one dislikes them, let him take the good, honest, ringing charge of Colonel Parker, " Remember that the whole boy goes to school ! "

Yes, the whole boy does go to school ; but the whole boy is seldom educated after he gets there. A fraction of him is attended to in the evening, however, and a fraction on Sunday. He takes himself in hand on Saturdays and in vacation time, and accomplishes a good deal, notwithstanding the fact that his sight is à trifle impaired already, and his hearing grown a little dull, so that Dame Nature works at a disadvantage, and begins, doubtless, to dread boys who have enjoyed too much " schooling," since it seems to leave them in a state of coma.

Our general scheme of education furthers mental development with considerable success. The training of the hand is now

being laboriously woven into it; but, even
when that is accomplished, we shall still be
working with imperfect aims, for the stress
laid upon heart-culture is as yet in no way
commensurate with its gravity.   We know,
with that indolent, fruitless half-knowledge
that passes for knowing, that "out of the
heart are the issues of life."   We feel, not
with the white heat of absolute conviction,
but placidly and indifferently, as becomes
the dwellers in a world of change, that
"conduct is three fourths of life;" but we
do not crystallize this belief into action.
We "dream," not "do" the "noble things."
The kindergarten does not fence off a half
hour each day for moral culture, but keeps
it in view every moment of every day.   Yet it
is never obtrusive; for the mental faculties
are being addressed at the same time, and
the body strengthened for its special work.

With the methods generally practiced in
the family and school, I fail to see how we
can expect any more delicate sense of right
and wrong, any clearer realization of duty,
any greater enlightenment of conscience, any
higher conception of truth, than we now
find in the world.   I care not what view you
take of humanity, whether you have Calvin-

istic tendencies and believe in the total depravity of infants, or whether you are a disciple of Wordsworth and apostrophize the child as a

> " Mighty prophet ! Seer blest,
>   On whom those truths do rest
> Which we are toiling all our lives to find ; "

if you are a fair-minded man or woman, and have had much experience with young children, you will be compelled to confess that they generally have a tolerably clear sense of right and wrong, needing only gentle guidance to choose the right when it is put before them. I say most, not all, children ; for some are poor, blurred human scrawls, blotted all over with the mistakes of other people. And how do we treat this natural sense of what is true and good, this willingness to choose good rather than evil, if it is made even the least bit comprehensible and attractive? In various ways, all equally dull, blind, and vicious. If we look at the downright ethical significance of the methods of training and discipline in many families and schools, we see that they are positively degrading. We appoint more and more "monitors" instead of training the "inward monitor" in each child, make truth-telling difficult instead of easy, punish trivial and grave offenses about in the same way, practice open bribery by promising children a few cents a day to behave themselves, and weaken their sense of right by giving them picture cards for telling the truth and credits for doing the most obvious duty. This has been carried on until we are on the point of needing another Deluge and a new start.

Is it strange that we find the moral sense blunted, the conscience unenlightened? The moral climate with which we surround the child is so hazy that the spiritual vision grows dimmer and dimmer, — and small wonder! Upon this solid mass of ignorance and stupidity it is difficult to make any impression; yet I suppose there is greater joy in heaven over a cordial "thwack" at it than over most blows at existing evils.

The kindergarten attempts a rational, respectful treatment of children, leading them to do right as much as possible for right's sake, abjuring all rewards save the pleasure of working for others and the delight that follows a good action, and all punishments save those that follow as natural penalties, of broken laws, — the obvious consequences

of the special bit of wrong-doing, whatever
it may be. The child's will is addressed in
such a way as to draw it on, if right; to
turn it willingly, if wrong. Coercion in the
sense of fear, personal magnetism, nay, even
the child's love for the teacher, may be used
in such a way as to weaken his moral force.
With every free, conscious choice of right, a
human being's moral power and strength of
character increase; and the converse of this
is equally true.

If the child is unruly in play, he leaves
the circle and sits or stands by himself, a
miserable, lonely unit until he feels again in
sympathy with the community. If he destroys his work, he unites the tattered fragments as best he may, and takes the moral
object lesson home with him. If he has
neglected his own work, he is not given the
joy of working for others. If he does not
work in harmony with his companions, a
time is chosen when he will feel the sense of
isolation that comes from not living in unity
with the prevailing spirit of good will. He
can have as much liberty as is consistent
with the liberty of other people, but no
more. If we could infuse the *spirit* of this
kind of discipline into family and school

life, making it systematic and continuous from the earliest years, there would be fewer morally "slack-twisted" little creatures growing up into inefficient, bloodless manhood and womanhood. It would be a good deal of trouble; but then, life is a good deal of trouble anyway, if you come to that. We cannot expect to swallow the universe like a pill, and travel on through the world " like smiling images pushed from behind."

Blind obedience to authority is not in itself moral. It is necessary as a part of government. It is necessary in order that we may save children dangers of which they know nothing. It is valuable also as a habit. But I should never try to teach it by the story of that inspired idiot, the boy who " stood on the burning deck, whence all but him had fled," and from whence he would have fled if his mental endowment had been that of ordinary boys. For obedience must not be allowed to destroy common sense and the feeling of personal responsibility for one's own actions. Our task is to train responsible, self-directing agents, not to make soldiers.

Virtue thrives in a bracing moral atmosphere, where good actions are taken rather as a matter of course. The attempt to instill an idea of self-government into the tiny slips of humanity that find their way into the kindergarten is useful, and infinitely to be preferred to the most implicit obedience to arbitrary command. In the one case, we may hope to have, some time or other, an enlightened will and conscience struggling after the right, failing often, but rising superior to failure, because of an ever stronger joy in right and shame for wrong. In the other, we have a "*good goose*," who does the right for the picture card that is set before him, — a "trained dog" sort of child, who will not leap through the hoop unless he sees the whip or the lump of sugar. So much for the training of the sense of right and wrong! Now for the provision which the kindergarten makes for the growth of certain practical virtues, much needed in the world, but touched upon all too lightly in family and school.

The student of political economy sees clearly enough the need of greater thrift and frugality in the nation; but where and when do we propose to develop these virtues? Precious little time is given to them in most schools, for their cultivation does not

yet seem to be insisted upon as an integral
part of the scheme.   Here and there an inspired human being seizes on the thought
that the child should really be taught how
to live at some time between the ages of six
and sixteen, or he may not learn so easily
afterward.   Accordingly, the pupils under
the guidance of that particular person catch
a glimpse of eternal verities between the
printed lines of their geographies and gram-
mars.   The kindergarten makes the growth
of every-day virtues so simple, so gradual,
even so easy, that you are almost beguiled
into thinking them commonplace.   They
seem to come in, just by the way, as it were,
so that at the end of the day you have seen
thought and word and deed so sweetly min-
gled that you marvel at the "universal
dovetailedness of things," as Dickens puts
it.   They will flourish better in the school,
too, when the cheerful hum of labor is heard
there for a little while each day.   The kindergarten child has "just enough" strips
for his weaving mat, — none to lose, none to
destroy; just enough blocks in each of his
boxes, and every one of them, he finds, is
required to build each simple form.   He cuts
his square of paper into a dozen crystal-

shaped bits, and behold ! each one of these tiny flakes is needed to make a symmetrical figure. He has been careless in following directions, and his form of folded paper does not "come out" right. It is not even, and it is not beautiful. The false step in the beginning has perpetuated itself in each succeeding one, until at the end either partial success or complete failure meets his eye. How easy here to see the relation of cause to effect! "Courage!" says the kindergartner; " better fortune next time, for we will take greater pains." "Can you rub out the ugly, wrong creases?" "We will try. Alas, no! Wrong things are not so easily rubbed out, are they?" "Use your worsted quite to the end, dear : it costs money." "Let us save all the crumbs from our lunch for the birds, children ; do not drop any on the floor: it will only make work for somebody else." And so on, to the end of the busy, happy day. How easy it is in the kindergarten, how seemingly difficult later on ! It seems to be only books afterward; and "books are good enough in their own way, but they are a mighty bloodless substitute for life."

The most superficial observer values the

industrial side of the kindergarten, because it falls directly in line with the present effort to make some manual training a part of school work ; but twenty or twenty-five years ago, when the subject was not so popular, kindergarten children were working away at their pretty, useful tasks, — tiny missionaries helping to show the way to a truth now fully recognized. As to the value of leading children to habits of industry as early in life as may be, that they may see the dignity and nobleness of labor, and conceive of their individual responsibilities in this world of action, that is too obvious to dwell upon at this time.

To Froebel, life, action, and knowledge were the three notes of one harmonious chord ; but he did not advocate manual training merely that children might be kept busy, nor even that technical skill might be acquired. The piece of finished kindergarten work is only a symbol of something more valuable which the child has acquired in doing it.

The first steps in all the kindergarten occupations are directed or suggested by the teacher ; but these dictations or suggestions are merely intended to serve as a sort of

staff, by which the child can steady himself
until he can walk alone. It is always the
creative instinct that is to be reached and
vivified : everything else is secondary. By
reproduction from memory of a dictated
form, by taking from or adding to it, by
changing its centre, corners, or sides, — by
a dozen ingenious preliminary steps, — the
child's inventive faculty is developed ; and
he soon reaches a point in drawing, building, modeling, or what not, where his greatest delight is to put his individual ideas into
visible shape. The simple request, " Make
something pretty of your own," brings a
score of original combinations and designs,
— either the old thoughts in different shape
or something fresh and audacious which
hints of genius. Instead of twenty hack-
neyed and slavish copies of one pattern, we
have twenty free, individual productions,
each the expression of the child's inmost
personal thought. This invests labor with a
beauty and power, and confers upon it a
dignity, to be gained in no other way. It
makes every task, however lowly, a joy, because all the higher faculties are brought into
action. Much so-called " busy work," where
pupils of the " A class " are allowed to stick

a thousand pegs in a thousand holes while the "B class" is reciting arithmetic, is quite fruitless, because it has so little thought behind it.

Unless we have a care, manual training, when we have succeeded in getting it into the school, may become as mechanical and unprofitable as much of our mind training has been, and its moral value thus largely missed. The only way to prevent it is to borrow a suggestion from Froebel. Then, and only then, shall we have insight with power of action, knowledge with practice, practice with the stamp of individuality. Then doing will blossom into being, and "Being is the mother of all the little doings as well as of the grown-up deeds and heroic sacrifices."

The kindergarten succeeds in getting these interesting and valuable free productions from children of four or five years only by developing, in every possible way, the sense of beauty and harmony and order. We know that people assume, somewhat at least, the color of their surroundings ; and, if the sense of beauty is to grow, we must give it something to feed upon.

The kindergarten tries to provide a room,

more or less attractive, quantities of pictures and objects of interest, growing plants and vines, vases of flowers, and plenty of light, air, and sunshine. A canary chirps in one corner, perhaps; and very likely there will be a cat curled up somewhere, or a forlorn dog which has followed the children into this safe shelter. It is a pretty, pleasant, domestic interior, charming and grateful to the senses. The kindergartner looks as if she were glad to be there, and the children are generally smiling. Everybody seems alive. The work, lying cosily about, is neat, artistic, and suggestive. The children pass out of their seats to the cheerful sound of music, and are presently joining in an ideal sort of game, where, in place of the mawkish sentimentality of "Sally Walker," of obnoxious memory, we see all sorts of healthful, poetic, childlike fancies woven into song. Rudeness is, for the most part, banished. The little human butterflies and bees and birds flit hither and thither in the circle; the make-believe trees hold up their branches, and the flowers their cups; and everybody seems merry and content. As they pass out the door, good-bys and bows and kisses are wafted backward into the room; for the manners of polite society are observed in everything.

You draw a deep breath. This is a *real* kindergarten, and it is like a little piece of the millennium. " Everything is so very pretty and charming," says the visitor. Yes, so it is. But all this color, beauty, grace, symmetry, daintiness, delicacy, and refinement, though it seems to address and develop the æsthetic side of the child's nature, has in reality a very profound ethical significance. We have all seen the preternatural virtue of the child who wears her best dress, hat, and shoes on the same august occasion. Children are tidier and more careful in a dainty, well-kept room. They treat pretty materials more respectfully than ugly ones. They are inclined to be ashamed, at least in a slight degree, of uncleanliness, vulgarity, and brutality, when they see them in broad contrast with beauty and harmony and order. For the most part, they try " to live up to " the place in which they find themselves. There is some connection between manners and morals. It is very elusive and, perhaps, not very deep; but it exists. Vice does not flourish alike in all conditions and localities, by any means. An ignorant negro was overheard praying, " Let me so lib dat when I die I may *hab manners*, dat I may know what to say when I see my heabenly Lord! " Well, I dare say we shall need good manners as well as good morals in heaven ; and the constant cultivation of the one from right motives might give us an unexpected impetus toward the other. If the systematic development of the sense of beauty and order has an ethical significance, so has the happy atmosphere of the kindergarten an influence in the same direction.

I have known one or two " solid men " and one or two predestinate spinsters who said that they did n't believe children could accomplish anything in the kindergarten, because they had too good a time. There is something uniquely vicious about people who care nothing for children's happiness. That sense of the solemnity of mortal conditions which has been indelibly impressed upon us by our Puritan ancestors comes soon enough, Heaven knows! Meanwhile, a happy childhood is an unspeakably precious memory. We look back upon it and refresh our tired hearts with the vision when experience has cast a shadow over the full joy of living.

The sunshiny atmosphere of a good kindergarten gives the young human plants an impulse toward eager, vigorous growth. Love's warmth surrounds them on every side, wooing their sweetest possibilities into life. Roots take a firmer grasp, buds form, and flowers bloom where, under more unfriendly conditions, bare stalks or pale leaves would greet the eye, — pathetic, unfulfilled promises, — souls no happier for having lived in the world, the world no happier because of their living. "Virtue kindles at the touch of joy." The kindergarten takes this for one of its texts, and does not breed that dismal fungus of the mind "which disposes one to believe that the pursuit of knowledge must necessarily be disagreeable."

The social phase of the kindergarten is most interesting to the student of social economics. Coöperative work is strongly emphasized; and the child is inspired both to live his own *full* life, and yet to feel that his life touches other lives at every point, — "for we are members one of another." It is not the unity of the "little birds," in the couplet, who "agree" in their "little nests," because "they 'd fall out if they did n't," but a realization, in embryo, of the divine principle that no man liveth to himself.

As to specifically religious culture, everything fosters the spirit out of which true religion grows.

In the morning talks, when the children are most susceptible and ready to "be good," as they say, their thoughts are led to the beauty of the world about them, the pleasure of right doing, the sweetness of kind thoughts and actions, the loveliness of truth, patience, and helpfulness, and the goodness of the Creator to all created things. No parent, of whatever creed or lack of creed, whether a bigot or unbeliever, could object to the kind of religious instruction given in the kindergarten; and yet in every possible way the child-soul and the child-heart are directed towards everything that is pure and holy, true and steadfast.

If the child love not his brother whom he hath seen, how can he love God whom he hath not seen? "Love worketh no ill to his neighbor, therefore love is the fulfilling of the law." There is a vast deal of practical religion to be breathed into these little children of the street before the abstractions of beliefs can be comprehended. They cannot live on words and prayers and texts, the thought and feeling must come before the expression. As Mrs. Whitney says, "The world is determined to vaccinate children with religion for fear they should take it in the natural way."

Some wise sayings of the good Dr. Holland, in "Nicholas Minturn," come to me as I write. Nicholas says, in discussing this matter of charities, and the various means of effecting a radical cure of pauperism, rather than its continual alleviation : "If you read the parable of the Sower, I think that you will find that soil is quite as necessary as seed — indeed, that the seed is thrown away unless a soil is prepared in advance. . . . I believe in religion, but before I undertake to plant it, I would like something to plant it in. The sowers are too few, and the seed is too precious to be thrown away and lost among the thorns and stones."

Last, but by no means least, the admirable physical culture that goes on in the kindergarten is all in the right direction. Physiologists know as much about morality as ministers of the gospel. The vices which drag men and women into crime spring as often from unhealthy bodies as from weak wills and callous consciences. Vile fancies and sensual appetites grow stronger and more terrible when a feeble physique and low vitality offer no opposing force. Deadly vices are nourished in the weak, diseased bodies that are penned, day after day, in filthy, crowded tenements of great cities. If we could withdraw every three-year-old child from these physically enfeebling and morally brutalizing influences, and give them three or four hours a day of sunshine, fresh air, and healthy physical exercise, we should be doing humanity an inestimable service, even if we attempted nothing more.

I have tried, as briefly as I might in justice to the subject, to emphasize the following points : —

I. That we must act up to our convictions with regard to the value of preventive work. If we are ever obliged to choose, let us save the children.

II. That the relation of the kindergarten to social reform is simply that, as a plan of education, it offers us valuable suggestions in regard to the mental, moral, and physical culture of children, which, in view of certain crying evils of the day, we should do well to follow.

The essential features of the kindergarten

which bear a special relation to the subject
are as follows : —

1. The symmetrical development of the
child's powers, considering him neither as
all mind, all soul, nor all body; but as a
creature capable of devout feeling, clear
thinking, noble doing.

2. The attempt to make so-called " moral
culture " a little less immoral; the rational
method of discipline, looking to the growth
of moral, self-directing power in the child,
— the only proper discipline for future citizens of a free republic.

3. The development of certain practical
virtues, the lack of which is endangering
the prosperity of the nation; namely, economy, thrift, temperance, self-reliance, frugality, industry, courtesy, and all the sober
host, — none of them drawing-room accom-
plishments, and consequently in small demand.

4. The emphasis placed upon manual
training, especially in its development of the
child's creative activity.

5. The training of the sense of beauty,
harmony, and order; its ethical as well as
æsthetical significance.

6. The insistence upon the moral effect

of happiness; joy the favorable climate of childhood.

7. The training of the child's social nature; an attempt to teach the brotherhood of man as well as the Fatherhood of God.

8. The realization that a healthy body has almost as great an influence on morals as a pure mind.

I do not say that the consistent practice of these principles will bring the millennium in the twinkling of an eye, but I do affirm that they are the thought-germs of that better education which shall prepare humanity for the new earth over which shall arch the new heaven.

Ruskin says, " Crime can only be truly hindered by letting no man grow up a criminal, by taking away the will to commit sin ! " But, you object, that is sheer impossibility. It does seem so, I confess, and yet, unless you are willing to think that the whole plan of an Omnipotent Being is to be utterly overthrown, set aside, thwarted, then you must believe this ideal possible, somehow, sometime.

I know of no better way to grow towards it than by living up to the kindergarten idea, that just as we gain intellectual power by doing intellectual work, and the finest æsthetic feeling by creating beauty, so shall we win for ourselves the power of feeling nobly and willing nobly by doing " noble things."

HOW SHALL WE GOVERN OUR CHILDREN?

"Not the cry," says a Chinese author, "but the rising of a wild duck, impels the flock to follow him in upward flight."

HOW SHALL WE GOVERN OUR CHILDREN?

LONG ago, in a far-off country, a child was born; and when his parents looked on him they loved him, and they resolved in their simple hearts to make of him a strong, brave, warlike man. But the God of that country was a hungry and an insatiable God, and he cried out for human sacrifice; so, when his arms had been thrice heated till they glowed red with the flame of the fire, the mother cradled her child in them, and his life exhaled as a vapor.

A child was born in another country, and the tender eyes of his mother saw that his limbs were misshapen and his life-blood a sickly current. Yet her heart yearned over him, and she would have tended and trained him and loved him better than all the rest of her strong, well-favored brood; but when the elders of her people knew that the child was a weakling, they decreed that he should die, and she bent her head to the law, which was stronger than her love.

In a third land a child was to be born, and the proud father made ready gifts, and purchased silken robes, and prepared a feast for his friends ; but, alas! when the longed-for soul entered the world it was housed in a woman-child's body, and straightway the joy was changed into mourning. Bitter reproaches were heaped upon the mother, for were there not enough women already on the earth? and the fiat went forth that the babe should straightway be delivered from the trials of existence. So, while its hold on life was yet uncertain, the husband's mother placed wet cloths upon its lips, and soon the faint breath stopped, and the white soul went fluttering heavenward again.

In still another of God's fair lands a child entered the world, and he grew toward manhood vigorous and lusty ; but he heeded not his parents' commands, and when his disobedience had been long continued, the fathers of the tribe decreed that he should be stoned to death, for so it was written in the sacred books. And as the youth was the absolute property of his parents, and as by common consent they had full liberty to deal with him as seemed good to them, they consented unto his death, that his soul might be saved alive, and the evening sun shone crimson on his dead body as it lay upon the sands of the desert.

At a later day and in a Christian country two children were born, one hundred years apart, and the world had now so far progressed that absolute power over the life of the offspring was denied the parents. The one was ruled with iron rods; he was made to obey with a rigidity of compliance and a severity of treatment in case of failure which made obedience a slavish duty, and he was taught besides that he was a child of Satan and an heir of hell. He found no joy in his youth, and his miserable soul groveled in fear of the despot who dominated him, and of the blazing eternity which he was told would be the punishment for his sins. His will was broken; he was made weak where he might have been strong; and he did evil because he had learned no power of self-restraint: yet his people loved him, and they had done all these things because they wished to purge him wholly from all uncleanness.

The parents of the other child were warned of the lamentable results of this gloomy training, and they said one to another: "Our darling shall be free as air; his duties shall be made to seem like pleasures, or, better still, he shall have no duty but his pleasure. He shall do only what he wills, that his will may grow strong, and he can but choose the right, for he knows no evil. We will hold up before him no bugbear of future punishment, for doubtless there is no such thing; and if there be, it will not be meted out to such a child. He will love and obey his parents because they have devoted themselves to his happiness, and because they have never imposed distasteful obligations upon him, and when he grows to manhood he will be a model of wisdom and of goodness."

But, lo! the child of this training was as great a failure as the child of austerity and gloom. He was capricious, lawless, willful, disobedient, passionate; he thought of no one's pleasure save his own; he cared for his parents only in so far as they could be of use to him; and like a wild beast of the jungle he preyed upon the life around him, and cared not whom he destroyed if his appetites were satisfied.

" In every field of opinion and action, men are found swinging from one extreme to the other of life's manifold arcs of vibration." This perpetual movement may be the essential condition of existence, for death is cessation of motion ; or it may be a never-ending effort of the mind to reach an ideal which discloses itself so seldom as to make its permanent abiding-place a matter of uncertainty. Doubtless there is somewhere a middle to the arc, and in the lapse of ages the needle may at last find the " pole-point of central truth " and be at rest; but as yet, in every department of labor and thought, it is vibrating, and after tarrying a while at one extreme it swings unsatisfied back to the other.

Nowhere are these extremes more noticeable than in the government of children. Centuries ago, in the patriarchal period, the father of the family seems also to have exercised the functions of a criminal judge; but the uniting of the two sets of duties in one person does not appear to have inspired the children with insurmountable awe, for laws are found both in Numbers and Deuteronomy fixing the penalty of disobedience, and of the striking of a parent by a child.

Still later, the Roman father possessed

arbitrary powers of life and death over his children ; but it is probable that natural affection and a more advanced civilization commonly made the law a dead letter.

Though the world in time grew to feel that life belonged to the being who held it, not to those who gave it birth, still discipline has for ages been directed more to the body than to the mind, with an idea apparently that the pains of the flesh will save the soul.   Pious parents until within recent dates have regarded the flogging of children as absolutely a religious obligation, and many a tender mother has steeled her heart and strengthened her arm to give the blows which she regarded as essential to the spiritual well-being of her child.

The birch rod and the Bible were the Parents' Complete Guide to domestic management in Puritan days, and no one can deny that this treatment, . though rather a heroic one, seems to have produced fine, strong, self-denying men and women.

Governor Bradford, in 1648, speaks feelingly of the godliness of a Puritan woman whose office it was to " sit in a convenient place in the congregation, with a little birchen rod in her hand, and keep the children in great awe ; " and, from the frequency with which chastisement is mentioned in early Puritan records, it seems pretty clear that the sober little lads and lasses of the day did not suffer from over-indulgence.

When this wholesale whipping began to fall into disuse, many philosophers prophesied the ruin of the race, but these gloomy predictions have scarcely found their fulfillment as yet.

There has been, however, a colossal change in discipline, from the days when disobedience was punishable with death to the agreeable moral suasion of the nineteenth century, as exemplified in the " fin de siècle " nonsense rhyme : —

> " There once was a hopeful young horse
> Who was brought up on love, without force :
> He had his own way, and they sugared his hay ;
> So he never was naughty, of course."

The results of this delightful method of treatment seem rather problematic, and the modern child is universally acknowledged to be no improvement upon his predecessors in point of respect and filial piety at least.

A superintendent's report, written thirty years ago for one of the New England States, regrets that, even then, home government

had grown lax.   He wittily says that Young
America is *rampant*, parental influence
couchant ; and no reversal of these posi-
tions is as yet visible in 1892.

To those who note the methods by which
many children are managed, it is a matter
of wonderment that the results in character
and conduct are not very much worse than
they are.   Dr. Channing wisely says, " The
hope of the world lies in the fact that parents cannot make of their children what
they will."   Happy accidents of association
and circumstance sometimes nullify the
harm the parent has done, and the tremendous momentum of the race-tendency carries the child over many an obstacle which
his training has set in his path.

It seems crystal-clear at the outset that
you cannot govern a child if you have never
learned to govern yourself.   Plato said, many
centuries ago : " The best way of training
the young is to train yourself at the same
time ; not to admonish them, but to be
always carrying out your own principles in
practice," and all the wisdom of the ancients
is in the thought.   If, then, you are a fit
person to be trusted with the government of
a child, what goal do you propose to reach

in your discipline; what is your aim, your ideal?

1. The discipline should be thoroughly in harmony with child-nature in general, and suited to the age and development of the particular child in question.

2. It should appeal to the higher motives, and to the higher motives alone.

3. It should develop kindness, helpfulness, and sympathy.

4. It should never use weapons which would tend to lower the child's self-respect.

5. It should be thoroughly just, and the punishment, or rather the retribution, should be commensurate with the offense.

6. It should teach respect for law, and for the rights of others.

Finally, it should teach " voluntary obedience, the last lesson in life, the choral song which rises from all elements and all angels," and, as the object of true discipline is the formation of character, it should produce a human being master of his impulses, his passions, and his will.

The journey's end being fixed, one must next decide what route will reach it, and will be short, safe, economical, and desirable; and the roads to the presumably ideal discipline are many and well-traveled. Some of them, it is true, lead you into a swamp, some to the edge of a precipice ; some will hurl you down a mountain-side with terrific rapidity ; others stop half-way, bringing you face to face with a blank wall ; and others again will lose you entirely on a bleak and trackless plain. But no matter which route you select, you will have the wise company of a great many teachers, parents, and guardians, and an innumerable throng of fair and lovely children will journey by your side.

The road of threat and fear, of arbitrary and over-severe punishment, has been much traveled in all times, though perhaps it is a little grass-grown now.

The child who obeys you merely because he fears punishment is a slave who cowers under the lash of the despot. Undue severity makes him a liar and a coward. He hates his master, he hates the thing he is made to do ; there is a bitter sense of injustice, a seething passion of revenge, forever within him ; and were he strong enough he would rise and destroy the power that has crushed him. He has done right because he was forced to do so, not because he desired it ; and since the right-doing, the obedience, was neither the fruit of his reason nor his love, it cannot be permanent.

The feeling of justice is strong in the child's mind, and you have constantly wounded that feeling. You have destroyed the sense of cause and effect by your arbitrary punishments. You have corrected him for disobedience, for carelessness, for unkindness, for untruthfulness, for noisiness, and for slowness in learning his lessons.

How is he to know which of these offenses is the greatest, if all have received the same punishment? Why should giving him a good thrashing teach him to be kind to his little sister? Why should he learn the multiplication table with greater rapidity because you ferule him soundly? Have you ever found pain an assistance to the memory?

If he has little intellectual perception of the difference between truth and falsehood, why should you suppose that smart strokes on any portion of the body would quicken that perception?

Is it not clear as the sun at noonday that, since he observes the punishment to have no necessary relation to the offense, and since he observes it to be light or severe according to your pleasure, — is it not clear that he will suppose you to be using your superior strength in order to treat him unfairly, and will not the supposition sow seeds of hatred and rebellion in his heart?

Another road to discipline is that of bribery.

To endeavor to secure goodness in a child by means of bribery, to promise him a reward in case he obeys you, is manifestly an absurdity. You are destroying the very traits in his character you are presumably endeavoring to build up. You are educating a human being who knows good from evil, and who should be taught deliberately to choose the right for the right's sake, who should do his duty because he knows it to be his duty, not for any extraneous reward connected with it. A spiritual reward will follow, nevertheless, in the feeling of happiness engendered, and the child may early be led to find his satisfaction in this, and in the approval of those he loves.

There are, of course, certain simple rewards which can be used with safety, and which the child easily sees to be the natural results of good conduct. If his treatment of the household pussy has been kind and gentle, he may well be trusted with a pet of his own; if he puts his toys away carefully when asked to do so, father will notice the neat room when he comes home; if he learns his lessons well and quickly, he will have the more time to work in the garden; and the suggestion of these natural consequences is legitimate and of good effect.

It is always safer, no doubt, to appeal to a love of pleasure in children than to a fear of pain, yet bribes and extraneous rewards inevitably breed selfishness and corruption, and lead the child to expect conditions in life which will never be realized. Though retribution of one kind or another follows quickly on the heels of wrong-doing, yet virtue is commonly its own reward, and it is as well that the child should learn this at the beginning of life. Froebel says: " Does a simple, natural child, when acting rightly, think of any other reward which he might receive for his action than this consciousness, though that reward be only praise? . . .

" How we degrade and lower the human nature which we should raise, how we weaken those whom we should strengthen, when we hold up to them an inducement to act virtuously ! "

Emulation is often harnessed into service

to further intellectual progress and the formation of right habits of conduct, and this inevitably breeds serious evils.

It is well to set before the child an ideal on which he may form himself as far as possible; but when this ideal sits across the aisle, plays in a neighboring back yard, or, worse still, is another child in the same family, he is hated and despised. His virtues become obnoxious, and the unfortunate evildoer prefers to be vicious, that he may not resemble a creature whose praises have so continually been sung that his very name is odious.

If the child grows accustomed to the comparison of himself with others and the endeavor to excel them, he becomes selfish, envious, and either vain of his virtue and attainments, or else thoroughly disheartened at his small success, while he grudges that of his neighbor. George Macdonald says: "No work noble or lastingly good can come of emulation, any more than of greed. I think the motives are spiritually the same."

To what can we appeal, then, in children, as motives to goodness, as aids in the formation of right habits of thought and action? Ah! the child's heart is a harp of many strings, and touched by the hand of a master a fine, clear tone will sound from every one of them, while the resultant strain will be a triumphant burst of glorious harmony.

Touch delicately the string of love of approval, and listen to the answer.

The child delights to work for you, to please you if he can, to do his tasks well enough to win your favorable notice, and the breath of praise is sweet to his nostrils. It is right and justifiable that he should have this praise, and it will be an aid to his spiritual development, if bestowed with discrimination. Only Titanic strength of character can endure constant discouragement and failure, and yet work steadily onward, and the weak, undeveloped human being needs a word of approval now and then to show him that he is on the right track, and that his efforts are appreciated. Of course the kind and the frequency of the praise bestowed depend entirely upon the nature of the child.

One timid, self-distrustful temperament needs frequently to bask in the sunshine of your approval, while another, somewhat predisposed to vanity and self-consciousness, needs a more bracing moral climate.

There is no question that cleanliness and fresh air may be considered as minor aids to goodness, and a dangerous outbreak of insubordination may sometimes be averted by hastily suggesting to the little rebel a run in the garden, prefaced by a thorough application of cool water to the flushed face and little clenched hands ; while self-respect may often be restored by the donning of a clean apron.

Beauty of surroundings is another incentive to harmony of action. It is easier for the child to be naughty in a poor, gloomy room, scanty of furniture, than in a garden gay with flowers, shaded by full-leafed trees, and made musical by the voice of running water.

Dr. William T. Harris says : " Beauty cannot create a new heart, but it can greatly change the disposition," and this seems unquestionable, especially with regard to the glory of God's handiwork, which makes goodness seem " the natural way of living." Yet we would not wish our children to be sybarites, and we must endeavor to cultivate in their breasts a hardy plant of virtue which will live, if need be, on Alpine heights and feed on scanty fare.

It is a truism that interesting occupation

prevents dissension, and that idle fingers are the Devil's tools.

A child who is good and happy during school time, with its regular hours and alternated work and play, often becomes, in vacation, fretful, sulky, discontented, and in arms against the entire world.

The discipline of work, if of a proper kind, of a kind in which success is not too long delayed, is sure and efficacious. Success, if the fruit of one's own efforts, is so sweet that one longs for more of the work which produced it.

The reverse of the medal may be seen here also. The knotted thread which breaks if pulled too impatiently; the dropped stitches that make rough, uneven places in the pattern; the sail which was wrongly placed and will not propel the boat; the pile of withered leaves which was not removed, and which the wind scattered over the garden, — are not all these concrete moral lessons in patience, accuracy, and carefulness?

We may safely appeal to public opinion, sometimes, in dealing with children. The chief object in doing this "is to create a constantly advancing ideal toward which the child is attracted, and thereby to gain a constantly increasing effort on his part to realize this ideal." There comes a time in the child's development when he begins to realize his own individuality, and longs to see it recognized by others. The views of life, the sentiments of the people about him, are clearly noted, and he desires to so shape his conduct as to be in harmony with them. If he sees that tale-bearing and cowardice are looked upon with disgust by his comrades, he will be a very Spartan in his laconicism and courage; if his father and older brothers can bear pain without wincing, then he will not cry when he hurts himself.

Oftentimes he is obdurate when reproved in private for a fault, but when brought to the tribunal of the disapproval of other children, he is chagrined, repents, and makes atonement. He is uneasy under the adverse verdict of a large company, but the condemnation of one person did not weigh with him. It is usually not wise, however, to appeal to public opinion in this way, save on an abstract question, as the child loses his self-respect, and becomes degraded in his own eyes, if his fault is trumpeted abroad.

Stories of brave deeds, poems of heroism, self-sacrifice, and loyalty, have their places in creating a sentiment of ideality in the child's breast, — a sentiment which remains fixed sometimes, even though it be not in harmony with the feeling of the majority.

Now and then some noble soul is born, some hero so thrilled with the ideal that he rises far above the public sentiment of his day; but usually we count him great who overtops his fellows by an inch or two, and he who falls much below the level of ordinary feeling is esteemed as almost beyond hope.

To seek for the approval of others, even though they embody our highest ideals, is truly not the loftiest form of aspiration; but it is one round in the ladder which leads to that higher feeling, the desire for the benediction of the spirit-principle within us.

Although discipline by means of fear, as the word is commonly used, cannot be too strongly condemned, yet there is a " godly fear " of which the Bible speaks, which certainly has its place among incentives in will-training. The child has not attained as yet, and it is doubtful whether we ourselves have done so, to that supreme excellence of love which absolutely casteth out fear.

A writer of great moral insight says: " Has not the law of seed and flower, cause and effect, the law of continuity which binds the universe together, a tone of severity? It has surely, like all righteous law, and carries with it a legitimate and wholesome fear. If we are to reap what we have sown, some, perhaps most of us, may dread the harvest."

The child shrinks from the disapproval of the loved parent or teacher. By so much the more as he reverences and respects those "in authority over him" does he dread to do that which he knows they would condemn. If he has been led to expect natural retributions, he will have a wholesome fear of putting his hand in the fire, since he knows the inevitable consequences. He understands that it is folly to expect that wrong can be done with impunity, and shrinks in terror from committing a sin whose consequences it is impossible that he should escape. He knows well that there are other punishments save those of the body, and he has felt the anguish which follows self-condemnation. "There is nothing degrading in such fear, but a heart-searching reverence and awe in the sincere and humble conviction that God's law is everywhere."

Such are some of the false and some of the true motives which can be appealed to in will-training, but there are various points in their practical application which may well be considered.

May we not question whether we are not frequently too exacting with children, — too much given to fault-finding? Were it not that the business of play is so engrossing to them, and life so fascinating a matter on the whole, — were it not for these qualifying circumstances, we should harass many of them into dark cynicism and misanthropy at a very early age. I marvel at the scrupulous exactness in regard to truth, the fine sense of distinction between right and wrong, which we require of an unfledged human being who would be puzzled to explain to us the difference between a " hawk and a handsaw," who lives in the realm of the imagination, and whose view of the world is that of a great play-house furnished for his benefit. If we were one half as punctilious and as hypercritical in our judgment of ourselves, we should be found guilty in short order, and sentenced to hard labor on a vast number of counts.

There are many comparatively small

faults in children which it is wise not to see at all. They are mere temporary failings, tiny drops which will evaporate if quietly left in the sunshine, but which, if opposed, will gather strength for a formidable current. If we would sometimes apply Tolstoi's doctrine of non-resistance to children, if we would overlook the small transgression and quietly supply another vent for the troublesome activity, there would be less clashing of wills, and less raising of an evil spirit, which gains wonderful strength while in action.

Do we not often use an arbitrary and a threatening manner in our commands to children, when a calm, gentle request, in a tone of expectant confidence, would gain obedience far more quickly and pleasantly?

Some natures are antagonized by the shadow of a threat, even if it accompanies a reasonable order; and if we acknowledge that the oil of courtesy is a valuable lubricator in our dealings with grown people, it seems proper to suppose that it would not be entirely useless with children. We cannot expect to get from them what we do not give ourselves, and it is idle to imagine that we can address them as we would a disobedient dog, and be answered in tones of dulcet harmony.

Again, what possible harm can there be in sometimes giving reasons for commands, when they are such as the child would appreciate? We do not desire to bring him up under martial rule; and if he feels the wisdom of the order issued, he will be much more likely to obey it pleasantly. Cases may frequently occur in which reasons either could not properly be given, or would be beyond the child's power of comprehension; but if our treatment of him has been uniformly frank and affectionate, he will cheerfully obey, believing that, as our commands have been reasonable heretofore, there is good cause to suppose they may still be so.

Educational opinion tends, more and more every day, to the absolute conviction that the natural punishment, the effect which follows the cause, is the only one which can safely be used with children.

This is the method of Nature, severe and unrelenting it may be, but calm, firm, and purely just. He who sows the wind must reap the whirlwind, and he who sows thistles may be well assured that he will never gather figs as his harvest. The feeling of continuity, of sequence, is naturally strong in the child; and if we would lead him to appreciate that the law is as absolute in the moral as in the physical world, we shall find the ground already prepared for our purpose.

Much transgression of moral law in later years is due to the fatal hope in the evil-doer's mind that he will be able to escape the consequences of his sin. Could we make it clear from the beginning of life that there is no such escape, that the mills of the gods will grind at last, though the hopper stand empty for many a year, — could we make this an absolute conviction of the mind, I am assured that it would greatly tend to lessen crime.

And this is one of the defects of arbitrary punishment, that it is sometimes withheld when the heart of the judge melts over the sinner, leading him to expect other possible exemptions in the future. Is it not sometimes given in anger, also, when the culprit clearly sees it to be disproportionate to the crime?

Here appears the advantage of the natural punishment, — it is never withheld in weak affection, it is never given in anger, it is entirely disassociated from personal feeling. No poisoned arrow of injustice remains rankling in the child's breast; no rebellious feeling that the parent has taken advantage of his superior strength to inflict the punishment: it is perceived to be absolutely *fair*, and, being fair, it must be, although painful, yet satisfactory to that sense of justice which is a passion of childhood.

Our American children are as precocious in will-power as they are keen-witted, and they need a special discipline. The courage, activity, and pioneer spirit of the fathers, exercised in hewing their way through virgin forests, hunting wild beasts in mountain solitudes, opening up undeveloped lands, prospecting for metals through trackless plains, choosing their own vocations, helping to govern their country, — all these things have reacted upon the children, and they are thoroughly independent, feeling the need of caring for themselves when hardly able to toddle.

Entrust this precocious bundle of nerves and individuality to a person of weak will or feeble intelligence, and the child promptly becomes his ruler.   The power of strong volition becomes caprice, he does not learn the habit of obedience, and thus valuable directive power is lost to the world.

" The lowest classes of society," says Dr. Harris, " are the lowest, not because there is any organized conspiracy to keep them down, but because they are lacking in directive power."   The jails, the prisons, the reformatories, are filled with men who are there because they were weak, more than because they were evil.   If the right discipline in home and school had been given them, they would never have become the charge of the nation.   Thus we waste force constantly, force of mind and of spirit sufficient to move mountains, because we do not insist that every child shall exercise his " inherited right," which is, " that he be taught to obey."

It is a grave subject, this of will-training, the gravest perhaps that we can consider, and its deepest waters lie far below the sounding of my plummet.   Some of the principles, however, on which it rests are as firmly fixed as the bed of the ocean, which remains changeless though the waves continually shift above : —

1. If we can but cultivate the *habit* of doing right, we enlist in our service one of the strongest of human agencies. Its momentum is so great that it may propel the child into the course of duty before he has time to discuss the question, or to parley with his conscience concerning it.

2. We must remember that "force of character is cumulative, and all the foregone days of virtue work their health into this." The task need not be begun afresh each morning ; yesterday's strokes are still there, and to-day's efforts will make the carving deeper and bolder.

3. We may compel the body to carry out an order, the fingers to perform a task ; but this is mere slavish compliance.' True obedience can never be enforced ; it is the fruit of the reason and the will, the free, glad offering of the spirit.

4. Though many motives have their place in early will-training, — love of approval, deference to public opinion, the influence of beauty, hopeful occupation, respect and reverence for those in authority, — yet these are all preparatory, the preliminary exercises, which must be well practiced before the soul can spread her wings into the blue.

5. There is but one true and final motive to good conduct, and that is a hunger in the soul of man for the blessing of the spirit, a ceaseless longing to be in perfect harmony with the principles of everlasting and eternal right.

THE MAGIC OF "TOGETHER"

" ' Together ' is the key-word of the nineteenth century.''

THE MAGIC OF "TOGETHER"

IT is an old, adobe-walled Mexican gar-den. All around it, close against the brown bricks, the fleur-de-lis stand white and stately, guarded by their tall green lances. The sun's rays are already powerful, though it is early spring, and I am glad to take my book under the shade of the orange-trees. In the dark leaf-canopy above me shine the delicate star-like flowers, the partly opened buds, and the great golden oranges, while tiny green and half-ripe spheres make a happy contrast in color. The ground about me is strewn with flowers and buds, the air is heavy with fragrance, and the bees are buzzing softly overhead. I am growing drowsy, but as I lift my eyes from my book they meet something which interests me. A large black ant is tugging and pulling at an orange-bud, and really making an effort to carry it away with him. It is once and a half as long as he, fully twice as wide, and I cannot compute how much heavier, but its size and weight are very little regarded. He

drags it vigorously over Alpine heights and
through valley deeps, but evidently finds the
task arduous, for he stops to rest now and
then.   I want to help him, but cannot be
sure of his destination, and fear besides that
my clumsy assistance would be misinter-
preted.

Ah, how unfortunate! ant and orange-bud
have fallen together into the depths of a
Colorado cañon which yawns in the path.
The ant soon reappears, but clearly feels it
impossible to drag the bud up such a precipice, and runs away on some. other quest.
What did he want with that bud, I wonder?
was it for food, or bric-a-brac, or a plaything
for the babies?   Never mind, — I shall
never know, and I prepare to read again.
But what's this? Here is my ant returning,
and accompanied by some friends.   They
disappear in the cañon, helpfulness and interest in every wave of their feelers.   Their
heads come into sight again, and — yes!
they have the bud.   Now, indeed, events
move, and the burden travels rapidly across
the smooth courtyard toward the house.
Can they intend to take it up on the flat
roof, where we have lately suspected a nest?
Yes, there they go, straight up the wall, all

putting their shoulders to the wheel, and resting now and then in the chinks of the crumbling adobes. Up the bud moves to the gutters, — I can see it gleam as it is pulled over the edge, — they are out of sight, — the task is done! How easy any undertaking, I think, when people are willing to help.

In a high dormer window of a great city, in a nest of quilts and pillows, sits little Ingrid. Her blue Danish eyes look out from a pinched, snow-white face, and her thin hands are languidly folded in her lap. She gazes far down below to the other side of the square, where she can just see the waving of some green branches and an open door.

Her eyes brighten now, for a stream of little children comes pouring from that door. " Look, mother! " she cries, " there are the children! " and the mother leaves her washing, and comes with dripping hands to see every tiny boy look up at the window and flourish his hat, and every girl wave her handkerchief, or kiss her hand. They form a ring ; there is silence for a moment and then, 'mid great flapping of dingy handkerchiefs and battered hats, a hearty cheer is heard.

"They're cheering my birthday," cries Ingrid. "Miss Mary knows it's my birthday. Oh, isn't it lovely!" And the thin hands eagerly waft some grateful kisses to the group below.

The scene has only lasted a few moments, the children have had their run in the fresh air, and now they go marching back, pausing at the door to wave good-by to the window far above. The mother carries Ingrid back to her bed (it is a weary time now since those little feet touched the floor); but the bed is not as tiresome as usual, nor the washing as hard, for both hearts are full of sunshine.

Afternoon comes, — little feet are heard climbing up the stair, and Ingrid's name is called. The door opens, and two flushed and breathless messengers stand on the threshold. "We've brung you a birfday present," they cry; "it's a book, and we made it all our own se'ves, and all the chilluns helped and made somefin' to put in it. Miss Mary's down stairs mindin' the babies, and she sends you her love. Good-by! Happy birfday!"

" Happy birthday " indeed ! Golden, precious, love-crowned birthday ! Was ever such a book, so full of sweet messages and tender thoughts !

Ingrid knows how baby Tim must have labored to sew that red circle, how John Jacob toiled over that weaving-mat, and Elsa carefully folded the drove of little pigs. Everybody thought of her, and all the "chilluns " helped, and how dear is the tangible outcome of the thoughts and the helping !

Far back in the childhood of the world, the long-haired savage, " woaded, winter-clad in skins," went roaming for his food wherever he might find it. He dug roots from the ground, he searched for berries and fruits, he hid behind rocks to leap upon his living prey, yet often went hungry to his lair at night, if the root-crop were short, or the wild beast wary.

But if the day had been a fortunate one, if his own stomach were filled and his body sheltered, little cared he whether long-haired savage number two were hungry and cold. " Every one for himself," would he say, as he rolled himself in his skins, " and the cave-bear, or any other handy beast, take the hindmost." The simplicity of his mental state, his complete freedom from responsibility, assure us that his digestion of the raw flesh and the tough roots must have been perfection, and the sleep in those furred skins a dreamless one.

What impending visitation of a common enemy, what sudden descent of a fierce horde of strange, wild, long-forgotten creatures, first moved him to ally himself with barbarians number two and three for their mutual protection ? And when long years of alliance in warfare, and mutual distrust at all other times, had slipped away, and when savages were turning into herdsmen and farmers and toolmakers, to what leader among men did a system of exchange of commodities for mutual convenience suggest itself ?

One would like to have met that painted savage who first suggested combination in warfare, or that later politico-economist upon whom it faintly dawned that mutual help was possible in other directions save that of blood-shedding.

A union born of the exigencies of warfare would be strengthened later by the promptings of self-interest, and, lo ! the experiment is no longer an experiment, and the fact is proven that men may fight and work together to their mutual profit and advancement.

'T is a simple proposition, after all, that ten times one is ten; and the bees, the ants, the grosbeaks, and the beavers prove it so clearly that any one of us may read, though we pass by never so quickly. Yet all great truths appear in man's mind in very rudimentary form at first, and each successive generation furnishes more favorable soil for their growth and development.

First, men joined hands in offensive and defensive alliance; second, they found that, even when wars were over, still communication, intercourse, and exchange of goods were desirable; third, they discovered that no great enterprise which would better their condition would be possible without coöperation; and, fourth, they began to band themselves together here and there, not only for their own protection, for their own gain, but to watch over the weak, to succor the defenseless, and even to uphold some dear belief.

The magic of "Together" has thus far reached, and who can tell what Happy Valley,

what fair Land of Beulah, it may summon
into existence in the future?

The incalculable value of coöperation,
the solemn truth that we are members one
of another, that we cannot labor for ourselves without laboring for others, nor injure
ourselves without injuring others, — all this
is intellectually appreciated by most men to-
day, all this is doubtless acknowledged; yet
I cannot find that it has obtained much recognition in education, nor is especially insisted upon in the training of children.

But surely, if children have any social
tendencies, — and the fact needs no proof, —
these tendencies should be given direction
from the beginning toward benevolence,
toward harmonious working together for
some common aim. This would be comparatively easy even in a nursery containing
three or four little people; and how much
simpler when school life begins, and when
the powers of children are greatly increased,
while they are in hourly contact with a
large number of equals!

"Society," as Dr. Hale says, "is the great
charm and only value of school life;" but
this charm and this value are reduced to
a minimum in many schools. "Emulation,

that devil-shadow of aspiration," so often used as a stimulus in education, must forever separate the child from his fellows.

How can I have any Christian fellowship with a man when I am envying him his successes and grudging him his honors? Am I not tempted to withhold my help from my weak brother across the way, lest my assistance place him on an equality with me?

Again, the "monitor" system, as sometimes carried out, tends to separation and engenders dislike and distrust. I am not likely to desire close communion, except in the way of fisticuffs, with a boy who has been spying upon me all day, or who has very likely "reported" me as having committed divers venial offenses.

It is the idea of some teachers that discipline is furthered if children are trained to have as little as possible to do with each other, and there is no question that this method does facilitate a toe-the-line kind of government. It would probably be more satisfactory to such a teacher if each child could be brought to school in a sedan-chair, with only one window and that in front, and could be kept in it during the whole session.

As such a plan, however, is scarcely feasible; as children, with or against our wills, have a natural and God-given instinct for each other's company; as they keenly enjoy banding themselves together for whatever purpose, should not education follow the suggestions which an earnest study of child-nature can but give?

Froebel, with those divinely curious eyes of his, saw deeper into the child's mind and heart than any of his predecessors, and for every faint stirring of life which he perceived provided adequate conditions of development. True prophet of the coming day, his philosophy is rich with suggestions for the cultivation of the social powers of the child. No one ever felt more keenly than he the inseparable, the organic connection of all life; and with deep spiritual insight he provides nursery plays and songs by which the babe, even in his mother's arms, may be led faintly to recognize in his being one of the links of the great chain which girdles the universe.

Later, when as a child of three or four years he makes his first step into the world, and loosing his mother's hand, enters a larger family of children of his own age, he is still led to feel himself a part of a vast union, each member of which has ministered to him, and numberless ways are opened by which he can join with others to give back to the world some of the benefits he has enjoyed. Stories are told and games are played which lead him to thank the kindly hands which have furnished his daily bread, his warm clothing, and his sweet, white bed at night.

The feeling of gratitude, grown and strengthened, must overflow in action. The world has done so much for him, what can he do for the world? Is there not some little invalid who would greatly prize a book of dainty pictures, embroidered, drawn, and painted by her child-friends? Then he will join with his companions, and patiently and lovingly fashion such a book. Is the class room somewhat bare and colorless? Then he can give up some of his cherished work to make a bright frieze about the walls.

A national holiday is perhaps approaching. He will unite with all the other babies in making flags, tri-colored chains, and rosettes to deck the room appropriately, and to please the mothers, fathers, and friends who are coming to celebrate the occasion.

One of the greatest pleasures which is offered is that of being allowed to " help " somebody. If a child is quick, neat, and careful, if he has finished his bit of work, he may go and help the babies, and very gently and very patiently he guides the chubby fingers, threads the needles, or ties on little caps, and conquers refractory buttons.

To be a " little helper," whether he is assisting his companions or the grown-up people about him, grows to seem the highest honor within his reach. He knows the joy of ministering unto others, and he feels that " to help is to do the work of the world."

Thus we endeavor to give external expression to the feelings stirring in the heart of the child, knowing that " even love can grow cold " if not nourished. The whole spirit of the work, if carried out as Froebel intended, must tend directly toward social evolution, and the intense personalism which is a distinguishing mark of our civilization, and is clearly seen in our children, needs anointing with the oil of altruism.

The circle in which the children stand for the singing is itself a perfect representation of unity. Hands are joined to make a "round and lovely ring." If any child is unkind, or regardless of the rights of others, it is easily seen that he not only makes himself unhappy, but seriously mars the pleasure of all the other children. If he willfully leaves the circle, a link in the chain is broken which can only be mended when he repents his folly and pleasantly returns to his place. Thus early he may be made to feel that all lives touch his own, and that his indulgence in selfish passion not only harms himself, but is the more blameworthy in that it injures others.

The songs and games cannot be happily carried on unless each child is not only willing to help, but willing also to give up his chief desires now and then. All the children would like to be the flowers in the garden, perhaps, but it is obvious that some must remain in the circle, in order that the fence be perfect, and prevent stray animals from destroying what we love and cherish. So there is constant surrendering of personal desires in recognition of the fact that others have equal rights, and that, after all, one part is as good as another, since all are essential to the whole.

In coöperative building, the children quickly see that the symmetrical figure which

four little ones have made together, uniting
their material, is infinitely larger and finer
than any one of them could have made alone.
If they are making a village at their little
tables, one builds the church, another work-
shops and stores, others schools and houses,
while the remainder make roads, lay out
gardens, plant trees, and plough the fields.
No one of the children had strength enough,
time enough, or material enough to build
the village alone, yet see how well and how
quickly it is done when we all help!

The sand-box, in which of course all
children delight, lends itself especially to
coöperative exercises. They gather around
it and plant gardens with the bright-colored
balls; they use it for geography, moulding
the hills, mountains, valleys, and tracing the
rivers near their homes; they arrange historical dramas, as " Paul Revere's Ride,"
or the " Landing of the Pilgrims : " but no
child does any one of these things alone;
there is constant and happy coöperation.

It is the aim of one day's exercise, perhaps, to retrace with the child the various
steps by which his comfortable chair and his
strong work-table have come to him.

(Across one end of the sand-box, a group
of children plant a forest with little pine

branches which they have brought. The wood-cutters come, fell the trees, and cut away the boughs. Another party of children bring the heavy teams, previously built from the play-material, harness in the horses (taken from a Noah's Ark), and prepare to carry off the logs. Now here come the road-makers, and they lay out a smooth, hard road for the teams, reaching to the very bank of the river, which another party of little ones has made. The logs are tumbled into the stream ; they float downward, are rafted, carried to the mill ; little sticks are furnished to represent the boards into which they are sawn ; and the lumber is taken to the cabinet-maker, that he may fashion our furniture.)

Though there be twenty children around the sand-box, yet all have been employed. Each has enjoyed his own work, yet appreciated the value of his neighbor's. They have worked together harmoniously and the doing has reacted upon the heart, and strengthened the feeling of unity which is growing within.

Such exercises cannot fail to teach the value and power of social effort, and the necessity of subordinating personal desires to the common good. Yet the development of individuality is not forgotten, for " our power as individuals depends upon our recognition of the rights of others."

It is true that the social problem is an intricate one and cannot be worked out, even partially, at any stage of education, unless the leader of the children be a true leader, and be enthusiastically convinced of the essential value of the principles on which the problem is based. Yet this might be said with equal truth of any educational aim, for the gospel must always have its interpreters, and some will ever give a more spiritual reading and seize the truth which was only half expressed, while others, dull-eyed, mechanical, " kill with the letter."

" After all," says Dr. Stanley Hall, " there is nothing so practical in education as the ideal, nor so ideal as the practical ; " and we may be assured that the direction of the social tendencies of the child toward high and noble aims, toward the sinking of self and the generous thought of others, — that this is not only ideal, not only a following after the purest light yet vouchsafed to us, but is at the same time practical in its detailed workings, and in its adaptation to the needs and desires of the day.

THE RELATION OF THE KINDER-GARTEN TO THE PUBLIC SCHOOL

"The nature of an educational system is determined by the manner in which it is begun."

THE RELATION OF THE KINDER-GARTEN TO THE PUBLIC SCHOOL

THE question for us to decide to-day is not how we can interest people in and how illustrate the true kindergarten, for that is already done to a considerable extent; but, how we can convince school boards, superintendents, and voters that the final introduction of the kindergarten into the public school system is a thing greatly to be desired. The kindergarten and the school, now two distinct, dissimilar, and sometimes, though of late very seldom, antagonistic institutions, — how will the one affect, or be affected by the other?

As to the final adoption of the kindergarten there is a preliminary question which goes straight to the root of the whole matter. At present the state accepts the responsibility of educating children after an arbitrarily fixed age has been reached. (Ought it not, rather, if it assumes the responsibility at all, to begin to educate the child when he *needs education?*)

Thoughtful people are now awaking to the fact that this regulation is an artificial, not a natural one, and that we have been wasting two precious years which might not only be put to valuable uses, but would so shape and influence after-teaching that every succeeding step would be taken with greater ease and profit. We have been discreet in omitting the beginning, so long as we did not feel sure how to begin. But we know now that Froebel's method of dealing with four or five year old babies, when used by a discreet and intelligent person, justifies us in taking this delicate, debatable ground.

So far, then, it is a question of law — a law which can be modified just as soon and as sensibly as the people wish. Before, however, that modification can become the active wish of the people, its importance must be understood and its effects estimated. Could it be shown that after-education will be hindered or in any way rendered more difficult by the kindergarten, clearly all efforts to introduce it must cease. Were it merely a matter of indifference, something that would neither make nor mar the after-work of schools, then it would remain a matter of choice or fancy, for individual parents to decide as they like; but, if it can be shown that the work of the kindergarten will lay a more solid foundation, or trace more direct paths for the workers of a later period, then it behooves us to give it a hearty welcome, and to work out its principles with zealous good will: and " working out " its principles means, *not* accepting it as a finality — a piece of flawless perfection — but as a stepping-stone which will lead us nearer to the truth. If it is a good thing, it is good for all; if it is truth, we want it everywhere; but if this new department of education and training is to gain ground, or accomplish the successful fruition of its wishes, there must be perfect unity among teachers concerning it. If they all understood the thing itself, and understood each other, there could be no lack of sympathy; yet there has been misunderstanding, conflict occasionally, and some otherwise worthy teachers have used the kindergarten as a sort of intellectual cuttle-fish to sharpen their conversational bills upon.

Of course I am not blind to the fact that after we have determined that we ought to have the kindergarten, there are many questions of expediency: suitable rooms, expense of material, salaries, assistants, age of children at entrance, system of government, number of children in one kindergarten; and greatest of all, but least thought of, strangely, the linking together of kindergarten and school, so that the development shall be continuous, and the chain of impressions perfect and unbroken.

Suffice it to say that it has been done, and can be done again; but it needs discretion, forethought, tact, earnestness, and unimpeachable honesty of administration, for unless we can depend upon our school boards and kindergartners *implicitly*, counting upon them for wise coöperation, brooding care, and great wisdom in selection of teachers, the experiment will be a failure. We have risks enough to run as it is; let us not permit our little ones, more susceptible by reason of age than any we have to deal with now, — let us not permit them to become victims of politics, rings, or machine teaching.

The kindergarten is more liable to abuse than any other department of teaching. There is no ground in the universe so sacred as this. But the difference between primary schools is just as great, only, unfortunately, we have become used to it; and the kindergarten being under fire, so to speak, must be absolutely ideal in its perfection, or it is ruthlessly held up to scorn.

There is a tremendous awakening all over the country with regard to kindergarten and primary work, and this is well, since the greatest and most fatal mistakes of the public school system have been made *just here;* and the time is surely coming when more knowledge, wisdom, tact, ingenuity, forethought, yes, and money, will be expended in order to meet the demands of the case. The time is coming when the imp of parsimony will no longer be mistaken for the spirit of economy ; when a woman possessed of ordinary human frailty will no longer be required to guide, direct, develop, train, help, love, and be patient with sixty little ones, just beginning to tread the difficult paths of learning, and each receiving just one sixtieth of what he craves. The millennium will be close at hand when we cease to expect from girls just out of the high school what Socrates never attempted, and would have deemed impossible.

Look at Senator Stanford's famous Palo Alto stock farm. Each colt born into that favored community is placed in a class of twelve. These twelve colts are cared for and taught by four or five trained teachers. No man interested in the training of fine horses ever objects, so far as I know, to such expenditure of labor and money. The end is supposed to justify the means. But when the creatures to be trained are human beings, and when the end to be reached is not race-horses, but merely citizens, we employ a very different process of reasoning.

That this subject of early training is a vitally interesting one to thinking people cannot be denied. The kindergarten has become the fashion, you say, cynically. This is scarcely true; but it is a fact that the upper, the middle, and the lower classes among us begin to recognize the existence of children under six years of age, and realize that far from being nonentities in life, or unknown quantities, they are very lively units in the sum of progressive education.

When we speak of kindergarten work among the children of the poor, and argue its claims as one of the best means of taking unfortunate little Arabs from the demoralizing life of the streets, and of giving their aimless hands something useful to do, their restless minds something good and

fruitful to think of, and their curious eyes
something beautiful to look on, there is not
a word of disapproval.   People seem willing
to concede its moral value when applied to
the lower classes, but, when they are obliged
to pay anything to procure this training for
their own children, or see any prospect of
what they call an already extravagant school
system made more so by its addition, they
become prolific in doubts.   In other words,
they believe in it when you call it *philan-
thropy*, but not when you call it *education ;*
and it must be called the germ of the better
education, toward which we are all struggling, the nearest approach to the perfect
beginning which we have yet found.

We see in the excellence of Froebel's idea,
educationally considered, its only claim to
peculiar power in dealing with incipient
hoodlumism.   It is only because it has such
unusual fitness to child-nature, such a store
of philosophy and ingenuity in its appli-
ances, and such a wealth of spiritual truth
in its aims and methods, that it is so great
a power with neglected children and ignorant
and vicious parents.

The principles on which Froebel built his
educational idea may be summed up briefly

under four heads. First, All the faculties of the child are to be drawn out and exercised as far as age allows. Second, The powers of habit and association, which are the great instruments of all education, of the whole training of life, must be developed with a systematic purpose from the earliest dawn of intelligence. Third, The active instincts of childhood are to be cultivated through manual exercise (chiefly creative in character), which is· made an essential part of the training, and this manual exercise is to be valued chiefly as a means of self-expression. Fourth, The senses are to be trained to accuracy as well as the hand. The child must learn how to observe what is placed before him, and to observe it truly, an acquirement which any teacher of science or art will appreciate. To work out these principles, Froebel devised his practical method of infant education, and the very name he gave to the place where his play lessons were to be used marks his purpose. No books are to be seen in a kindergarten, because no ideas or facts are presented to the child that he cannot clearly understand and verify. The object is not to teach him arithmetic or geometry, though he learns enough of both to be very useful to him hereafter; but to lead him to discover *truths* concerning forms and numbers, lines and angles, for himself.

Thus in the play-lessons the teacher simply rules the order in which the child shall approach a new thing, and gives him the correct names which, henceforth, he must always use; but the observation of resemblances and differences (that groundwork of all knowledge), the reasoning from one point to another, and the conclusions he arrives at, are all his own; he is only led to see his mistake if he makes one. The child handles every object from which he is taught, and learns to reproduce it.

It is not enough to say that any ordinary system of object teaching in the hands of an ingenious teacher will serve the purpose or take the place of the kindergarten. People who say this evidently have no conception of Froebel's plan, in which the simultaneous training of head, heart, and hand is the most striking characteristic.

The kindergarten is mainly distinguished from the later instruction of the school by making the knowledge of facts and the cultivation of the memory subordinate to the development of observation and to the appropriate activity of the child, physical, mental, and moral. Its aim is to utilize the now almost wasted time from four to six years, a time when all negligent and ignorant mothers leave the child to chance development, and when the most careful mother cannot train her child into the practice of social virtues so well as the truly wise kindergartner who works with her. "We learn through doing" is the watchword of the kindergarten, but it must be a *doing* which blossoms into *being*, or it does not fulfill its ideal, for it is character building which is to go on in the kindergarten, or it has missed Froebel's aim.

What does the kindergarten do for children under six years of age? What has it accomplished when it sends the child to the primary school? I do not mean what Froebel hoped could be done, or what is occasionally accomplished with bright children and a gifted teacher, or even what is done in good private kindergartens, for that is yet more; but I mean what is actually done for children by charitable organizations, which are really doing the work of the state.

I think they can claim tangible results

which are wholly remarkable; and yet they
do not work for results, or expect much
visible fruit in these tender years, from a
culture which is so natural, child-like, and
unobtrusive that its very outward simplicity has caused it to be regarded as a plaything.

In glancing over the acquirements of the
child who has left the kindergarten, and has
been actually *taught* nothing in the ordinary
acceptation of the word, we find that he has
worked, experimented, invented, compared,
reproduced. All things have been revealed
in the doing, and productive activity has enlightened and developed the mind.

First, as to arithmetic. It does not come
first, but though you speak with the tongues
of men and angels, and make not mention of
arithmetic, it profiteth you nothing. The
First Gift shows one object, and the children
get an idea of one whole; in the Second
they receive three whole objects again, but of
different form; in the Third and Fourth,
the regularly divided cube is seen, and all
possible combinations of numbers as far as
eight are made. In the Fifth Gift the child
sees three and its multiples; in fractions,
halves, quarters, eighths, thirds, ninths, and

twenty-sevenths. With the Sixth, Seventh, and Eighth Gifts the field is practically unlimited.

Second, as to the child's knowledge of form, size, and proportion. His development has been quite extensive : he knows, not always by name, but by their characteristics, vertical, horizontal, slanting, and curved lines ; squares, oblongs ; equal sided, blunt and sharp angled triangles ; five, six, seven and eight sided figures ; spheres, cylinders, cubes, and prisms. All this elementary geometry has, of course, been learned "baby fashion," in a purely experimental way, but nothing will have to be unlearned when the pupil approaches geometry later in a more thoroughly scientific spirit.

Third, as to the cultivation of language, of the power of expression, we cannot speak with too much emphasis. The vocabulary of the kindergarten child of the lower classes is probably greater than that of his mother or father. You can see how this comes about. The teachers themselves are obliged to make a study of simple, appropriate, expressive, and explicit language ; the child is led to express all his thoughts freely in proper words from the moment he can lisp ;

he is trained through singing to distinct and careful enunciation, and the result is a remarkably good power of language. I make haste to say that this need not necessarily be used for the purposes of chattering in the school.

The child has not, of course, learned to read and write, but reading is greatly simplified by his accurate power of observation, and his practice of comparing forms. The work of reading is play to a child whose eye has been thus trained. As to writing, we precede it by drawing, which is the sensible and natural plan. The child will have had a good deal of practice with slate and lead pencil; will have drawn all sorts of lines and figures from dictation, and have created numberless designs of his own.

If, in short, our children could spend two years in a good kindergarten, they would not only bring to the school those elements of knowledge which are required, but would have learned in some degree how to *learn,* and, in the measure of their progress, *have nothing to unlearn.*

Let those who labor, day by day, with inert minds never yet awakened to a wish for knowledge, a sense of beauty, or a feeling of

pleasure in mental activity, tell us how much
valuable school time they would save, if the
raw material were thus prepared to their
hand. "After spending five or six years
at home or in the street, without training or
discipline, the child is sent to school and is
expected to learn at once. He looks upon
the strange, new life with amazement, yet
without understanding. Finally, his mind
becomes familiar in a mechanical manner,
ill-suited to the tastes of a child, with the
work and exercises of primary instruction,
the consequence being, very often, a feeble
body and a stuffed mind, the stuffing having
very little more effect upon the intellect than
it has upon the organism of a roast turkey."
The kindergarten can remedy these intellectual difficulties, beside giving the child an
impulse toward moral self-direction, and a
capacity for working out his original ideas
in visible and permanent form, which will
make him almost a new creature. It can,
by taking the child in season, set the wheels
in motion, rouse all his best, finest, and highest instincts, the purest, noblest, and most
vivifying powers of which he is possessed.

There is a good deal of time spent in the
kindergarten on the cultivation of politeness

and courtesy; and in the entirely social atmosphere which is one of its principal features, the amenities of polite society can be better practiced than elsewhere.

The kindergarten aims in no way at making infant prodigies, but it aims successfully at putting the little child in possession of every faculty he is capable of using; at bringing him forward on lines he will never need to forsake; at teaching within his narrow range what he will never have to unlearn; and at giving him the wish to learn, and the power of teaching himself. Its deep simplicity should always be maintained, and no lover of childhood or thoughtful teacher would wish it otherwise. It is more important that it should be kept pure than that it should become popular.

I have tried, thus, somewhat at length, to demonstrate that our educational system cannot be perfect until we begin still earlier with the child, and begin in a more childlike manner, though, at the same time, earnestly and with definite purpose. In trying to make manhood and womanhood, we sometimes treat children as little men and women, not realizing that the most perfect childhood is the best basis for strong manhood.

Further, I have tried to show that Froebel's system gives us the only rational beginning ; but I confess frankly that to make it productive of its vaunted results, it must be placed in the hands of thoroughly trained kindergartners, fitted by nature and by education for their most delicate, exacting, and sacred profession.

Now as to compromises. The question is frequently asked, Cannot the best things of the kindergarten be introduced in the primary departments of the public school? The best thing of kindergartening is the kindergarten itself, and nothing else will do ; it would be necessary to make very material changes in the primary class which is to include a kindergarten — changes that are demanded by radically different methods.

The kindergarten should offer the child experience instead of instruction ; life instead of learning ; practical child-life, a miniature world, where he lives and grows, and learns and expands. No primary teacher, were she Minerva herself, can work out Froebel's idea successfully with sixty or seventy children under her sole care.

You will see for yourselves that this simple, natural, motherly instruction of babyhood cannot be transplanted bodily into the primary school, where the teacher has fifty or sixty children who are beyond the two most fruitful years which the kindergarten demands. Besides, the teachers of the lower grades cannot introduce more than an infinitesimal number of kindergarten exercises, and at the same time keep up their full routine of primary studies and exercises.

Any one who understands the double needs of the kindergarten and primary school cannot fail to see this matter correctly, and as I said before, we do not want a few kindergarten exercises, we want the *kindergarten*. If teachers were all indoctrinated with the spirit of Froebel's method, they would carry on its principles in dealing with pupils of any age; but Froebel's kindergarten, pure and simple, creates a place for children of four or five years, to begin their bit of life-work; it is in no sense a school, nor must become so, or it would lose its very essence and truest meaning.

Let me show you a kindergarten! It is no more interesting than a good school, but I want you to see the essential points of difference: —

It is a golden morning, a rare one in a

long, rainy winter.   As we turn into the narrow, quiet street from the broader, noisy
one, the sound of a bell warns us that we
are near the kindergarten building. . . . A
few belated youngsters are hurrying along,
— some ragged, some patched, some plainly
and neatly clothed, some finishing a " port-
able breakfast " thrust into their hands five
minutes before, but all eager to be there. . . .
While the Lilliputian armies are wending
their way from the yard to their various
rooms, we will enter the front door and look
about a little.

The windows are wide open at one end of
the great room.   The walls are tinted with
terra cotta, and the woodwork is painted in
Indian red.   Above the high wood dado
runs a row of illuminated pictures of animals, — ducks, pigeons, peacocks, calves,
lambs, colts, and almost everything else that
goes upon two or four feet ; so that the children can, by simply turning in their seats,
stroke the heads of their dumb friends of
the meadow and barnyard. . . . There are
a great quantity of bright and appropriate
pictures on the walls, three windows full of
plants, a canary chirping in a gilded cage,
a globe of gold-fish, an open piano, and an

old - fashioned sofa, which is at present adorned with a small scrap of a boy who clutches a large slate in one hand, and a mammoth lunch-pail in the other. . . . It is his first day, and he looks as if his big brother had told him that he would be " walloped " if he so much as winked.

A half-dozen charming girls are fluttering about ; charming, because, whether plain or beautiful, they all look happy, earnest, womanly, full to the brim of life.

> " A sweet, heart-lifting cheerfulness,
> Like spring-time of the year,
> Seems ever on their steps to wait."

. . . They are tying on white aprons and preparing the day's occupations, for they are a detachment of students from a kindergarten training school, and are on duty for the day.

One of them seats herself at the piano and plays a stirring march. The army enters, each tiny soldier with a "shining morning face." Unhappy homes are forgotten . . . smiles everywhere . . . everybody glad to see everybody else . . . happy children, happy teachers . . . sunshiny morning, sunshiny hearts . . . delightful work in prospect, merry play to follow it. . . . "Oh, it's a beautiful world, and I'm glad I'm in it;" so the bright faces seem to say.

It is a cosmopolitan regiment that marches into the free kindergartens of our large cities. Curly yellow hair and rosy cheeks . . . sleek blonde braids and calm blue eyes . . . swarthy faces and blue-black curls . . . woolly little pows and thick lips . . . long arched noses and broad flat ones. Here you see the fire and passion of the Southern races, and the self-poise, serenity and sturdiness of Northern nations. Pat is here with a gleam of humor in his eye . . . Topsy, all smiles and teeth, . . . Abraham, trading tops with Isaac, next in line, . . . Gretchen and Hans, phlegmatic and dependable, . . . François, never still for an instant, . . . Christina, rosy, calm, and conscientious, and Duncan, as canny and prudent as any of his people. Pietro is there, and Olaf, and little John Bull.

What an opportunity for amalgamation of races, and for laying the foundation of American citizenship! for the purely social atmosphere of the kindergarten makes it a life-school, where each tiny citizen has full liberty under the law of love, so long as he does not interfere with the liberty of his neighbor. The phrase " Every man for himself " is never heard, but " We are members one of another " is the common principle of action.

The circles are formed. Every pair of hands is folded, and bright eyes are tightly closed to keep out " the world, the flesh," and the rest of it, while children,and teachers sing one of the morning hymns : —

" Birds and bees and flowers,
    Every happy day,
Wake to greet the sunshine,
    Thankful for its ray.
All the night they 're silent,
    Sleeping safe and warm ;
God, who knows and loves them,
    Will keep them from all harm.

" So the little children,
    Sleeping all the night,
Wake with each new morning,
    Fresh and sweet and bright.
Thanking God their Father
    For his loving care,
With their songs and praises
    They make the day more fair."

Then comes a trio of good-morning songs, with cordial handshakes and scores of kisses wafted from finger-tips. . . . " Good-Morning, Merry Sunshine," follows, and the sun, encouraged by having some notice taken of him in this blind and stolid world, shines brighter than ever. . . . The song, "Thumbs and Fingers say 'Good-Morning,'" brings two thousand fingers fluttering in the air (10×200, if the sum seems too difficult), and gives the eagle-eyed kindergartners an opportunity to look for dirty paws and preach the needed sermon.

It is Benny's birthday; five years old to-day. He chooses the songs he likes best, and the children sing them with friendly energy. . . . "Three cheers for Benny, — only three, now!" says the kindergartner. . . . They are given with an enthusiasm that brings the neighbors to the windows, and Benny, bursting with pride, blushes to the roots of his hair. The children stop at three, however, and have let off a tremendous amount of steam in the operation. Any wholesome device which accomplishes this result is worthy of being perpetuated. . . . A draggled, forsaken little street-cat sneaks in the door, with a pitiful mew. (I'm sure I don't wonder! if one were tired of life, this would be just the place to take a fresh start.) The children break into the song, "I Love Little Pussy, Her Coat is so Warm," and the kindergartner asks the small boy with the great lunch pail if he would n't like to give the kitty a bit of something to eat. He complies with the utmost solemnity, thinking this the queerest community he ever saw. . . . A broken-winged pigeon appears on the window-sill and receives his morning crumb; and now a chord from the piano announces a change of programme. The children troop to their respective rooms fairly warmed through with happiness and good will. Such a pleasant morning start to some who have been "hustled" out of a bed that held several too many in the night, washed a trifle (perhaps!), and sent off without a kiss, with the echo of a sick mother's wails, or a father's oaths, ringing in their ears!

After a few minutes of cheerful preparation, all are busily at work. Two divisions have gone into tiny, "quiet rooms" to grapple with the intricacies of mathematical relations. A small boy, clad mostly in red woolen suspenders, and large, high-topped boots, is passing boxes of blocks. He is awkward and slow. The teacher could do it more quietly and more quickly, but the kindergarten is a school of experience where ease comes, by and by, as the lovely result of repeated practice. . . . We hear an informal talk on fractions, while the cube is divided into its component parts, and then see a building exercise " by direction."

In the other " quiet room " they are building a village, each child constructing, according to his own ideas, the part assigned him.   One of them starts a song, and they all join in —

> " Oh ! builders we would like to be,
>   So willing, skilled, and strong ;
> And while we work so cheerily,
>   The time will not seem long."

" If we all do our parts well, the whole is sure to be beautiful," says the teacher. " One rickety, badly made building will spoil our village. I 'm going to draw a blackboard picture of the children who live in the village.   Johnny, you have n't blocks enough for a good factory, and Jennie has n't enough for hers.   Why don't you club together and make a very large, fine one ? "

This working for a common purpose, yet with due respect for individuality, is a very important part of kindergarten ethics.   Thus each child learns to subordinate himself to the claims and needs of society without losing himself.   " No man liveth to himself " is the underlying principle of action.

Coming back to the main room we find one division weaving bright paper strips into a mat of contrasting color, and note that the occupation trains the sense of color and of number, and develops dexterity in both hands.

But what is this merry group doing in the farther corner? These are the babies, bless them! and they are modeling in clay. What an inspired version of pat-a-cake and mud pies is this! The sleeves are pushed up, showing a high-water mark of white arm joining little brown paws. What fun! They are modeling the seals at the Cliff House (for this chances to be a California kindergarten), and a couple of two-year-olds, who have strayed into this retreat, not because there was any room for them here, but because there wasn't any room for them anywhere else, are slapping their lumps of clay with all their might, and then rolling it into caterpillars and snakes. This last is not very educational, you say, but " virtue kindles at the touch of joy," and some lasting good must be born out of the rational happiness that surrounds even the youngest babies in the kindergarten.

The sand-table in this room represents an

Italian or Chinese vegetable garden. The children have rolled and leveled the surface and laid it off in square beds with walks between. The planting has been " make believe," — a different kind of seed in each bed ; but the children have named them all, and labeled the various plats with pieces of paper, fastened in cleft sticks. A gardener's house, made of blocks, ornaments one corner, and near it are his tools, — watering-pot, hoe, rake, spade, etc., all made in card-board modeling.

We now pass up-stairs. In one corner a family of twenty children are laying designs in shining rings of steel ; and as the graceful curves multiply beneath their clever fingers, the kindergartner is telling them a brief story of a little boy who made with these very rings a design for a beautiful " rose window," which was copied in stained glass and hung in a great stone church, of which his father was the architect. ·

Another group of children is folding, by dictation, a four-inch square of colored paper. The most perfect eye‑measure, as well as the most delicate touch, is needed here. Constant reference to the " sharp " angle, " blunt " angle, square corner and right angle, horizontal and vertical lines, show that the foundation is being laid for a future clear and practical knowledge of geometry, though the word itself is never mentioned.

There is one unhappy little boy in this class. He has broken the law in some way, and he has no work.

" That is a strange idea," said the woman visitor. " In my time work was given to us as a punishment, and it seemed a most excellent plan."

" We look at it in another way," said the kindergartner, smiling. " You see, work is really the great panacea, the best thing in the world. We are always trying to train the children to a love of industry and helpful occupation ; so we give work as a reward, and take it away as a punishment."

We pass into the sunny upper hall, and find some children surrounding a large sandtable. The exercise is just finished, and we gaze upon a miniature representation of the Cliff House embankment and curving road, a section of beach with people standing (wooden ladies and gentlemen from a Noah's Ark), a section of ocean, and a perfect Seal Rock made of clay.

" Run down-stairs, Timmy, please, and ask Miss Ellen if the seals are ready." . . . Timmy flies. . . .

Presently the babies troop up, each carrying a precious seal extended on two tiny hands or reposing in apron. They are all bursting with importance. . . . Of course, the small Jonah of the flock tumbles up the stairs, bumps his nose, and breaks his treasure. . . . There is an agonized wail. . . . " *I bust my seal !* " . . . Some one springs to the rescue. . . . The seal is patched, tears are dried, and harmony is restored. . . . The animals are piled on the rocks in realistic confusion, and another class comes out with twenty-five paper fishes to be arranged in the waves of sand.

Later on, the sound of a piano invites us to witness the kindergarten play-time.

Through kindergarten play the child comes to know the external world, the physical qualities of the objects which surround him, their motions, actions, and reactions upon each other, and the relations of these phenomena to himself; a knowledge which forms the basis of that which will be his permanent stock in life. The child's fancy is healthily fed by images from outer life, and his curiosity by new glimpses of knowledge from the world around him.

There are plays and plays! The ordinary unguided games of childhood are not to be confounded for an instant with the genuine kindergarten plays, which have a far deeper significance than is apparent to the superficial observer. "Take the simplest circle game; it illustrates the whole duty of a good citizen in a republic. Anybody can spoil it, yet nobody can play it alone; anybody can hinder its success, yet no one can get credit for making it succeed."

The play is over; the children march back to their seats, and settle themselves to another period of work, which will last until noon. We watch the bright faces, cheerful, friendly chatter, the busy figures hovering over pleasant tasks, and feel that it has been good to pass a morning in this republic of childhood.

I have given you but a tithe of the whole argument, the veriest bird's-eye view; neither is it romance; it is simple truth; and, that being the case, how can we afford to keep Froebel and his wonderful influence on childhood out of a system of free education which has for its aim the development of

a free, useful, liberty-loving, self-governing
people? It is too great a factor to be disregarded, and the coming years will prove
it so; for the value of such schools is no
longer a matter of theory; they have been
tested by experience, and have won favor
wherever they have been given a fair trial.
But how important a work they have to do
in our scheme of public education is clear
only when we consider the conditions which
our public schools must meet nowadays.

On the theory upon which the state under-
takes the education of its youth at all — the
necessity of preparing them for intelligent
citizenship — a community might better
economize, if economize it must, anywhere
else than on the beginning. An enormous
immigrant population is pressing upon us.
The kindergarten reaches this class with
great power, and increases the insufficient
education within the reach of the children
who must leave school for work at the age
of thirteen or fourteen. It increases it, too,
by a kind of training which the child gets
from no other schooling, and brings him
under influences which are no small addition
to the sum total of good in his life.

The entire pedagogical world watches

with interest the educational awakening of which the kindergarten has been the dawn. If people really want to make the experiment, if parents and tax-payers are anxious to have for their younger children what seems so beneficent a training, then let them accept no compromises, but, after taking the children at a proper age, see to it that they get pure kindergarten, true kindergarten, and *nothing* but kindergarten till they enter the primary school. Then they will be prepared for study, and begin it with infinite zest, because they comprehend its meaning. Having had that beautiful beginning, every later step will seem glad to the child ; he will not see knowledge " through a glass darkly, but face to face," in her most charming aspect.

OTHER PEOPLE'S CHILDREN

"Where is thy brother Abel?"

OTHER PEOPLE'S CHILDREN

WE will suppose, for the sake of argument, that the rights of our own children are secured; but though such security betokens an admirable state of affairs, it does not cover the whole ground; there are always the "other people's children." The still small voice is forever saying, "Where is thy brother Abel?"

There are many matters to be settled with regard to this brother Abel, and we differ considerably as to the exact degree of our responsibility towards him. Some people believe in giving him the full privileges of brotherhood, in sharing alike with him in every particular, and others insist that he is no brother of theirs at all. Let the nationalists and socialists, and all the other reformers, decide this vexed question as best they can, particularly with regard to the "grown-up" Abels. Meanwhile, there are a few sweet and wholesome services we can render to the brother Abels who are not big enough to be nationalists and socialists, nor strong enough to fight for their own rights.

Among these kindly offices to be rendered, these practical agencies for making Abel a happy, self-helpful, and consequently a better little brother, we may surely count the free kindergarten.

My mind convinces me that the kindergarten idea is true; not a perfect thing as yet, but something on the road to perfection, something full of vitality and power to grow; and my heart tells me that there is no more beautiful or encouraging work in the universe than this of taking hold of the unclaimed babies and giving them a bit of motherliness to remember. The Free Kindergarten is the mother of the motherless, the father of the fatherless; it is the great clean broom that sweeps the streets of its parentless or worse than parentless children, to the increased comfort of the children, and to the prodigious advantage of the street.

We are very much interested in the cleaning of city streets, and well we may be; but up to this day a larger number of men and women have concerned themselves actively about sweeping them of dust and dirt than of sweeping them free of these children. (If dirt is misplaced matter, then what do you call a child who sits eternally on the curbstones and in the gutters of our tenement-house districts?

I believe that since the great Teacher of humanity spoke those simple words of eternal tenderness that voiced the mother side of the divine nature, — " Suffer little children to come unto me, and forbid them not," — I believe that nothing more heartfelt, more effectual, has come ringing down to us through the centuries than Froebel's inspired and inspiring call, " Come! let us live with the children! "

This work *pays*, in the best and the highest sense as well as the most practical.

It is true, the kindergartner has the child in her care but three or four hours a day; it is true, in most instances, that the home influences are all against her; it is true that the very people for whom she is working do not always appreciate her efforts; it is true that in many cases the child has been " born wrong," and to accomplish any radical reform she ought to have begun with his grandfather; it is true she makes failures now and then, and has to leave the sorry task seemingly unperformed, giving into the

mighty hand of One who bringeth order out
of chaos that which her finite strength has
failed to compass.   She hears discouraging
words sometimes, but they do not make
a profound impression, when she sees the
weary yet beautiful days go by, bringing
with them hourly rewards greater than
speech can testify!

She sees homes changing slowly but surely
under her quiet influence, and that of those
home missionaries, the children themselves;
she gets love in full measure where she least
expected so radiant a flower to bloom; she
receives gratitude from some parents far beyond what she is conscious of deserving; she
sees the ancient and respectable dirt-devil
being driven from many of the homes where
he has reigned supreme for years; she sees
brutal punishments giving place to sweeter
methods and kinder treatment; and she is
too happy and too grateful, for·these and
more encouragements, to be disheartened by
any cynical dissertations·on the determination of the world to go wrong and the impossibility of preventing it.

It is easier, in my opinion, to raise money
for, and interest the general man or woman
in, the free kindergarten than in any other

single charity. It is always comparatively easy to convince people of a truth, but it is much easier to convince them of some truths than of others. If you wish to found a library, build a hospital, establish a diet-kitchen, open a bureau for woman's work, you are obliged to argue more or less; but if you want money for neglected children, you have generally only to state the case. Everybody agrees in the obvious propositions, " An ounce of prevention " — " As the twig is bent " — " The child is father to the man " — " Train up a child " — " A stitch in time " — " Prevention is better than cure " — " Where the lambs go the flocks will follow " — " It is easier to form than to reform," and so on *ad infinitum* — proverbs multiply. The advantages of preventive work are so palpable that as soon as you broach the matter you ought to find your case proved and judgment awarded to the plaintiff, before you open your lips to plead.

The whole matter is crystal clear; for happily, where the protection of children is concerned, there is not any free-trade side to the argument. We need the public kindergarten educationally as the vestibule to

our school work.   We need it as a philan-
thropic agent, leading the child gently into
right habits of thought, speech, and action
from the beginning.   We need it to help in
the absorption and amalgamation of our
foreign element ; for the social training, the
opportunity for coöperation, and the purely
republican form of government in the kindergarten make it of great value in the development of the citizen-virtues, as well as
those of the individual.

I cannot help thinking that if this side of
Froebel's educational idea were more insisted on throughout our common school
system, we should be making better citizens
and no worse scholars.

If we believe in the kindergarten, if we
wish it to become a part of our educational
system, we have only to let that belief —
that desire — crystallize into action ; but
we must not leave it for somebody else to
do.

It is clearly every mother's business and
father's business, — spinsters and bachelors
are not exempt, for they know not in what
hour they may be snatched from sweet liberty, and delivered into sweeter slavery.   It
is a lawyer's business, for though it will

make the world better, it will not do it soon enough to lessen litigation in his time. It is surely the doctor's business, and the minister's, and that of the business man. It is in fact everybody's business.

The beauty of this kindergarten subject is its kaleidoscopic character; it presents, like all truth, so many sides that you can give every one that which he likes or is fitted to receive. Take the aggressively self-made man who thinks our general scheme of education unprofitable, — show him the kindergarten plan of manual training. He rubs his hands. " Ah ! that 's common sense," he says. " I don't believe in your colleges — I never went to college ; you may count on me."

Give the man of æsthetic taste an idea of what the kindergarten does in developing the sense of beauty; show him in what way it is a primary art school.

Explain to the musician your feeling about the influence of music ; show the physical-culture people that in the kindergarten the body has an equal chance with mind and heart.

Tell the great-hearted man some sad incident related to you by one of your kindergartners, and as soon as he can see through his tears, show him your subscription book.

Give the woman who cannot reason (and there are such) an opportunity to feel. There is more than one way of imbibing truth, fortunately, and the brain is not the only avenue to knowledge.

Finally, take the utter skeptic into the kindergarten and let the children convert him. It commonly is a "him" by the way. The mother-heart of the universe is generally sound on this subject.

But getting money and opening kindergartens are not the only cares of a Kindergarten Association. At least there are other grave responsibilities which no other organization is so well fitted to assume. These are the persistent working upon school boards until they adopt the kindergarten, and, much more delicate and difficult, the protection of its interests after it is adopted; the opening of kindergartens in orphanages and refuges where they prove the most blessed instrumentality for good; the spreading of such clear knowledge and intelligent insight into the kindergarten as shall prevent it from deterioration; the insistence upon kindergartners properly trained by properly qualified training teachers; •the gentle mothering and inspiring and helping those kindergartners to realize their fair ideals (for Froebel's method is a growing thing, and she who does not grow with it is a hopeless failure); the proper equipment and furnishing of class-rooms so that the public may have good object-lessons before its eyes; the insistence upon the ultimate ideals of the method as well as upon details and technicalities, — that is, showing people its soul instead of forever rattling its dry bones. And when all is said and done, the heaviest of the work falls upon the kinder-gartner. That is why I am convinced that we should do everything that sympathy and honor and money can do to exalt the office, so that women of birth, breeding, culture, and genius shall gravitate to it. The kin-dergartner it is who, living with the chil-dren, can make her work an integral part of the neighborhood, the centre of its best life. She it is, often, who must hold husband to wife, and parent to child; she it is after all who must interpret the aims of the Asso-ciation, and translate its noble theories into practice. (Ay! and there's the rub.) She it is, who must harmonize great ideal principles with real and sometimes sorry conditions. A Kindergarten Association stands for certain things before the community. It is the kindergartner alone who can prove the truth, who can substantiate the argument, who can show the facts. There is no more difficult vocation in the universe, and no more honorable or sacred one. If a kindergartner is looked upon, or paid, or treated as a nursery maid, her ranks will gradually be recruited from that source.

The ideal teacher of little children is not born. We have to struggle on as best we can, without her. She would be born if we knew how to conceive her, how to cherish her. She needs the strength of Vulcan and the delicacy of Ariel; she needs a child's heart, a woman's heart, a mother's heart, in one; she needs clear judgment and ready sympathy, strength of will, equal elasticity, keen insight, oversight; the buoyancy of hope, the serenity of faith, the tenderness of patience. "The hope of the world lies in the children." When we are better mothers, when men are better fathers, there will be better children and a better world. The sooner we feel the value of beginnings, the sooner we realize that we can put bunglers and botchers anywhere else better than in nursery, kindergarten, or primary school (there are no three places in the universe so " big with Fate "), the sooner we shall arrive at better results.

I am afraid it is chiefly women's work. Of course men can be useful in many little ways ; such as giving money and getting other people to give it, in influencing legislation, interviewing school boards, securing buildings, presiding over meetings, and giving a general air of strength and solidity to the undertaking. But the chief plotting and planning and working out of details must be done by women. The male genius of humanity begets the ideas of which each century has need (at least it is so said, and I have never had the courage to deny it or the time to look it up) ; but the female genius, I am sure, has to work them out, and " to help is to do the work of the world."

If one can give money, if only a single subscription, let her give it ; if she can give time, let her give that ; if she has no time for absolute work, perhaps she has time for the right word spoken in due season ; failing all else, there is no woman alive, worthy the name, who cannot give a generous heart-throb, a warm hand-clasp, a sunny, helpful smile, a ready tear, to a cause that concerns itself with childhood, as a thank-offering for her own children, a pledge for those the hidden future may bring her, or a consolation for empty arms.

There is always time to do the thing that *ought* to be, that *must* be done, and for that matter who shall fix the limit to our powers of helpfulness? It is the unused pump that wheezes. If our bounty be dry, cross, and reluctant, it is because we do not continually summon and draw it out. But if, like the patriarch Jacob's, our well is deep, it cannot be exhausted. While we draw upon it, it draws upon the unspent springs, the hillsides, the clouds, the air, and the sea ; and the great source of power must itself suspend and be bankrupt before ours can fail.

The kindergarten is not for the poor child alone, a charity; neither is it for the rich child alone, a luxury, corrective, or antidote ; but the ideas of which it tries to be the expression are the proper atmosphere for every child.

It is a promise of health, happiness, and usefulness to many an unfortunate little waif, whose earthly inheritance is utter blackness, and whose moral blight can be outgrown and succeeded by a development of intelligence and love of virtue.

The child of poverty and vice has still within him, however overlaid by the sins of ancestry, a germ of good that is capable of growth, if reached in time. Let us stretch out a tender strong hand, and touching that poor germ of good lifting its feeble head in a wilderness of evil, help it to live and thrive and grow!

9 781017 519358